PROTEIN BIOSYNTHESIS

Titles published in the series:

*Antigen-presenting Cells
*Complement
DNA Replication
Enzyme Kinetics
Gene Structure and Transcription
Genetic Engineering
*Immune Recognition
*B Lymphocytes
*Lymphokines
Membrane Structure and Function
Molecular Basis of Inherited Disease
Protein Biosynthesis
Protein Engineering
Protein Targeting and Secretion
Regulation of Enzyme Activity
*The Thymus

*Published in association with the British Society for Immunology.

Series editors

David Rickwood

Department of Biology, University of Essex, Wivenhoe Park,
Colchester, Essex CO4 3SQ, UK

David Male

Institute of Psychiatry, De Crespigny Park, Denmark Hill,
London SE5 8AF, UK

PROTEIN BIOSYNTHESIS

H.R.V.Arnstein

Department of Biochemistry, King's College, University of London, Strand, London WC2R 2LS, UK

R.A.Cox

The National Institute for Medical Research, Mill Hill, London NW7 1AA, UK

OXFORD UNIVERSITY PRESS

Oxford is a trade mark of Oxford University Press

Published in the United States
by Oxford University Press, New York

A catalogue record for this book is available from the British Library

Library of Congress Cataloging in Publication Data
Protein biosynthesis / H.R.V. Arnstein and R.A. Cox.
(In focus)
Includes bibliographical references and index.
1. Proteins—Synthesis.   I. Cox, R. A.   II. Title.   III. Series:
In focus (Oxford, England)
[DNLM: 1. Amino Acids.   2. Amino Acyl tRNA Synthetases.
3. Proteins—biosynthesis.   4. Ribosomes.   5. RNA, Messenger.
6. RNA, Transfer.   QU 55 A767p]
QP551.A693   1992   574.19'296—dc20   91-3653
ISBN 0-19-963040-2 (pbk.)

In Focus is a registered trade mark of the Chancellor, Masters, and
Scholars of the University of Oxford trading as Oxford University Press.

Typeset and printed by Information Press Ltd, Oxford, England.

To Ruth, Isobel, and Jane

Preface

Protein biosynthesis is of central importance for all living systems and the topic has aroused intense interest for over half a century. The assembly of polypeptide chains from amino acids and their subsequent modifications leading to the final three-dimensional protein structure are exceptionally complex processes that involve many components and utilize much of the cell's energy.

The published literature on the subject is vast and in a book of this size it would have been an impossible task to discuss even the most important features in great detail. We have decided, therefore, to present a general overview of the mechanisms of the translation of the genetic message and the subsequent stages in protein synthesis, together with a somewhat more detailed account of what we believe are particularly significant new developments. Even so, we have omitted details of the processing of mRNA transcripts although this process plays a central role in the biosynthesis of antibodies.

It is our hope that the book will be of interest to students, non-specialists, and others with little in-depth knowledge of the subject and that it will enable them to acquire information about both the fundamentals and recent advances in this fascinating field.

H.R.V.Arnstein
R.A.Cox

Acknowledgements

Where appropriate, the source of each illustration is given in the legend to the Figure and we gratefully acknowledge permission of both authors and publishers to use these illustrations.

We are particularly indebted to the following colleagues who provided us with original photographs: Dr M. Boublik, Roche Institute of Molecular Biology, Roche Research Center, Nutley, NJ 07110, USA (*Figure 3.10*); Dr R. Brimacombe, Max-Planck-Institut für Molekulare Genetik, Ihnestrasse 73, 1000 Berlin 13, Germany (*Figure 3.9 d* and *e*); Dr A.E. Dahlberg, Brown University, Providence, RI 02912, USA (*Figure 3.1*); Dr J. Frank, Wadsworth Center for Laboratories and Research, New York State Health Department, Albany, NY 12201, USA (*Figures 3.12* and *3.13*); Dr J. Lake, University of California at Los Angeles, Los Angeles, CA 90024, USA (*Figures 3.4, 3.9 a – c* and *3.11*); Dr P. Moore, Yale University, New Haven, CT 06511, USA (*Figure 3.3*); and Dr R.J. Planta, Vrije Universiteit de Boelelaan, 1081 HV Amsterdam, The Netherlands (*Figure 3.6*).

We also thank our colleagues who gave us permission to use their published work, as follows: Professor A.R. Fersht, Cambridge University Chemical Laboratory, Lensfield Road, Cambridge CB2 1EW (*Figure 2.4*); Dr R. Garrett, Universitets Institut for Biologisk Kemi B, Copenhagen, Denmark (*Figure 3.8*); Dr R.J.A. Grand, CRC Laboratories, Department of Cancer Studies, The University of Birmingham, Medical School, Birmingham B15 2TJ (*Figure 5.4*); Dr O. Hayaishi, Osaka Medical College, Takatsuki, Osaka 569, Japan (*Figure 5.6*); Dr S. Kornfeld, Department of Internal Medicine and Biochemistry, Washington University School of Medicine, St Louis, Missouri 63110, USA (*Figure 5.3*); Dr A. Liljas, Lund University, Sweden (*Figure 1.2*); Dr H. Noller, University of California at Santa Cruz, CA 95064, USA (*Figures 3.5* and *3.7*); Dr M. Nomura, University of California at Irvine, CA 92717, USA (*Figure 3.2*); Professor R.E. Rhoads, Department of Biochemistry, University of Kentucky, College of Medicine, 800 Rose Street, Lexington, KY 40536 – 0084, USA (*Figure 4.4b*); Professor A. Rich, Department of Biology, Massachusetts Institute of Technology, Cambridge, MA 02139, USA (*Figure 2.2*); Dr P. Schimmel, Department of Biology, Massachusetts Institute of Technology, Cambridge, MA 02139, USA (*Figures 2.5* and *2.6*); Dr I. Tinoco, University of California at Berkeley, Berkeley, CA 94720, USA (*Figure 1.3b*).

We thank the copyright holders for permission to use material, as follows: Academic Press (*Figures 3.3, 3.8, 3.10, 3.11, 3.12 a* and *b*); American Association for the Advancement of Science (*Figures 3.2*, copyright 1973, and *3.5*, copyright 1989); American Society for

Microbiology (*Figures 3.4, 3.6, 3.9 a – c* and *f – h*); The Biochemical Society and Portland Press (*Figure 5.4*); Cell Press (*Figures 3.1* and *3.7*); Cold Spring Harbor Laboratory Press (*Figure 1.2*); Elsevier Science Publishers (*Figure 3.9 d* and *e*); Elsevier Trends Journals (*Figure 4.4b*); FASEB Journal (*Table 5.2*); ICSU Press (*Figure 1.3b*); VCH Verlagsgesellschaft (*Figure 2.4*). *Figure 2.2* has been reproduced, with permission, from the *Annual Review of Biochemistry, 45* (1976), *Figures 2.5* and *2.6* have been reprinted from *Nature, 333*, p.140 and *337*, p. 479, respectively, copyright © 1988 and 1989 Macmillan Magazines Ltd, *Figure 3.12 c* and *d* from the *Biophysical Journal, 55*, 455 – 64 (1989) by copyright permission of the Biophysical Society, and *Figures 5.3* and *5.6*, with permission, from the *Annual Review of Biochemistry, 54* (1985).

We thank Simon A. Cox (Imperial Cancer Research Fund Laboratories, Clare Hall, South Mimms, Herts EN6 3LD) for critically reading the manuscript and for his helpful suggestions.

Contents

Abbreviations and units

Å	ångström unit $= 10^{-10}$m (0.1 nm)
A, P, and E sites	ribosomal binding sites for aminoacyl-tRNA, peptidyl-tRNA and exiting deacylated tRNA, respectively.
Da	dalton, unit of molecular mass ($\frac{1}{12}$ of $C = 1$).
gp	glycoprotein; a following number indicates the molecular mass in kDa. Thus, gp82 denotes a glycoprotein of M_r 82 000.
k_{cat}	catalytic constant (*also called* turnover number) = maximum or limiting reaction rate of an enzyme divided by the enzyme concentration.
K_M	Michaelis constant = substrate concentration at which the rate of an enzyme-catalysed reaction is half the maximum rate.
M_r	molecular mass relative to $\frac{1}{12}$ of the atomic mass ^{12}C.
p	protein; e.g. p78 denotes a protein of M_r 78 000 (*see* gp).

1

Introduction

---

1. The importance of proteins

'In the protein molecule Nature has devised a unique instrument in which an underlying simplicity is used to express great subtlety and versatility; it is impossible to see molecular biology in proper perspective until this peculiar combination of virtues has been clearly grasped.'

Francis Crick (1)

The world around us, both living and inanimate, in all its diversity is built up from fewer than one hundred elements. The living world comprises single cells or assemblies of cells. Cells are highly ordered as shown by their morphology and functions, and by their capacity to reproduce themselves precisely. The fertilized human embryo, which is approximately 0.1 mm in size, carries the body plan of the adult and also the timetable for the stages of development leading to maturity. Despite their differences, cells show a large measure of similarity in their biochemical processes. Biological reactions taking place within the cell occur at moderate temperatures in moderately concentrated salt solutions near neutral pH. The reactions are precisely controlled to an extent that is rarely matched by non-enzymic reactions. This precision is achieved by catalysis; these catalysts are termed enzymes, which are usually proteins or proteins associated with non-protein components essential for catalytic activity. In addition, cellular proteins have a structural function as components of the cytoskeleton and extracellular matrix, which are essential for the organization of cells and tissues. In both respects proteins are uniquely important.

2. Properties of amino acids and proteins

Proteins are polypeptides which comprise a precise number of amino acids present in a unique sequence and joined together by peptide (amide) bonds formed by condensation between the carboxyl group of one residue and the amino group of another with the elimination of water. Polypeptides are polymers of 20 different common amino acids (see *Table 1.1*) which (apart from glycine which is symmetrical) have the L-configuration. A particular protein may comprise one or more linear polypeptide chains. In each case the exact sequence of amino acids in the polypeptide chain is precisely determined by the appropriate gene. All (or almost all) amino acids are found in all proteins. Proteins related by function are also related by amino acid sequence. A comparison of the amino acid sequence between, for example, cytochrome *c* of different species reveals similarities in amino acid sequence with relatively few differences. The similarities indicate a common function; the differences reflect the influence of evolutionary pressures.

Table 1.1. Properties of the common amino acids

Amino acid	Abbreviations or symbols		M_r of residue at pH 7	Relative abundance in *E.coli* (%)	pK value of side chain	$\triangle G$ values for transfer of side chain from ethanol to water[++] (kcal/mole)
	three letters	one letter				
aliphatic						
alanine	Ala	A	71	13.0		0.75
valine	Val	V	99	6.0		1.70
leucine	Leu	L	113	7.8		2.4
isoleucine	Ile	I	113	4.4		2.95
glycine	Gly	G	57	7.8		0.00
proline	Pro	P	97	4.6		2.60
cysteine	Cys	C	103	1.8	8.3	1.00
methionine	Met	M	131	3.8		1.30
aromatic						
histidine	His	H	137	0.7	6.0	ND
phenylalanine	Phe	F	147	3.3		2.65
tyrosine	Tyr	Y	163	2.2	10.1	2.85
tryptophan	Trp	W	186	1.0		3.0
uncharged polar						
asparagine	Asn	N	114	9.9[+]		ND
glutamine	Gln	Q	128	10.8[*]		ND
serine	Ser	S	87	6.0		ND
threonine	Thr	T	101	4.6		0.45
charged polar						
lysine	Lys	K	129	7.0	10.5	1.50
arginine	Arg	R	157	5.3	12.5	0.75
aspartate	Asp	D	114	9.9[+]	3.9	ND
glutamate	Glu	E	128	10.8[*]	4.3	ND

[+]Total for aspartic acid and asparagine.
[*]Total for glutamic acid and glutamine.
ND No data available.
[++]The change in free energy measures the preference of the amino acid side chain for ethanol (a non-polar solvent) compared with water (a polar solvent). Thus tryptophan has the most and arginine the least hydrophobic side-chain.

Although proteins are essentially linear polymers they derive their biological activity from the capacity of the polypeptide chain to fold into precise three-dimensional structures stabilized by non-covalent bonds such as ionic bonds and hydrogen bonds formed between $C=O$ and $-NH-$ groups of the peptide bond (see *Figure 1.1*), and by the formation of intramolecular bonds through the

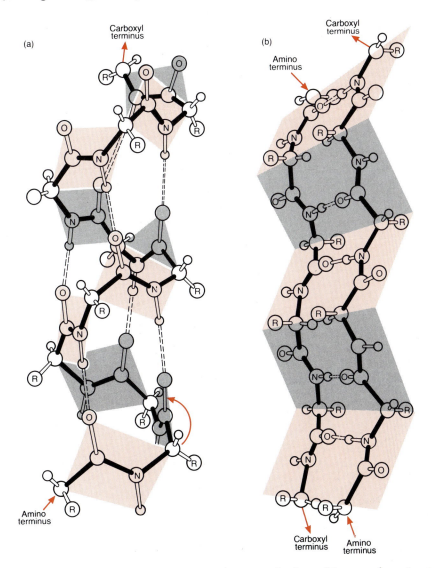

Figure 1.1. Two of the most common elements of polypeptide secondary structure. The polypeptide backbone (side-chains are omitted) is shown with a parallelogram highlighting the plane of each bond. (**a**) The right-handed α-helix. (**b**) Part of a β-sheet showing two component antiparallel strands. Note that alternate side-chains will be placed on opposite sides of the sheet. The polarity of the polypeptide chain is shown by the arrow which points from the amino terminus towards the carboxyl terminus.

formation of disulphide bridges. The precise three-dimensional structure (see for example carbonic anhydrase, *Figure 1.2*) is determined by the amino acid sequence of the polypeptide. The amino acids differ in their side chains, some of which are charged at the cellular pH of approximately pH 7.5 (see *Table 1.1*) so that, as with other polyelectrolytes, changing the pH of the environment may alter both structure and function. As shown in *Table 1.1*, some side-chains are polar (hydrophilic) and others are non-polar (hydrophobic). It is the wide range of noncovalent interactions which confers both versatility and subtlety upon protein structure. After the polypeptide chain has been synthesized it may be further modified by chemical reactions and by the addition of prosthetic groups such as haem or other low molecular weight compounds, or by chemical reactions involving amino acid side chains. Such modifications of amino acids are discussed in Chapter 5. Frequently, proteins comprise more than one polypeptide chain (see *Table 1.2*) whose amino acid sequences are encoded in different genes which in the case of eukaryotes are not necessarily located within the same chromosome. Alternatively, such multimeric proteins may arise by the proteolytic cleavage of a single-polypeptide precursor into two or more fragments.

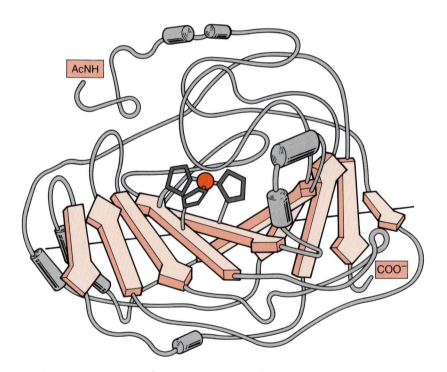

Figure 1.2. A schematic drawing of the tertiary structure of human carbonic anhydrase. The cylinders denote right-handed α helices (see *Figure 1.1a*) and the arrows denote individual strands of β-sheets (see *Figure 1.1b*) running from the amino to the carboxyl terminus. The structure was determined by X-ray crystallography (2). The orange circle in the centre represents Zn^{2+}.

Table 1.2. Subunit composition of proteins with more than one polypeptide chain (3)

Protein	Source	M_r of protein	No. of subunits	M_r of subunits
Insulin		11 466	2	5 733
Nerve growth factor	Mouse	26 518	2	13 259
Luteinizing hormone	Sheep	27 322	1	12 500
			1	14 830
Haemoglobin	Mammals	64 500	2	16 000
			2	16 000
Tu–Ts complex	E.coli	65 000	1	41 500
			1	28 500
Seryl tRNA synthetase	E.coli	100 000	2	50 000
lac repressor	E.coli	160 000	4	40 000
Methionyl tRNA synthetase	E.coli	170 000	2	85 000
Q_β replicase	E.coli	205 000	1	70 000
			1	65 000
			1	45 000
			1	35 000
RNA polymerase core	E.coli	400 000	2	39 000
			1	155 000
			1	165 000
Apoferritin	Horse	443 000	24	18 500
Ovomacroglobulin	Chicken	650 000	2	325 000
Pyruvate dehydrogenase complex	E.coli	5 000 000	24	91 000
			24	65 000
			24	56 000

3. The problem of protein biosynthesis

The problem of protein biosynthesis is essentially one of storage and retrieval of the genetic information which specifies the amino acid sequence of each of the hundreds of proteins present even in the smallest cell. For a polypeptide chain of 150 amino acids, any one of 20 amino acids may be present at any one point in the polymer. The chance of synthesizing the correct protein by reacting together 20 amino acids randomly is $(0.05)^{150}$ or 5×10^{-300}. Thus, for a hypothetical bacterium with a complement of, say, 1000 proteins each of 150 amino acids, the sequence information which must be stored in order to synthesize precisely each of the proteins when required is formidable. Handing on this information from one generation to the next without generating errors is part of the same problem.

The mechanism of protein biosynthesis is central to all molecular biology. Before the problem of protein biosynthesis was solved, researchers were confronted with the paradox that enzymes are the catalysts of the cell and the biosynthesis of proteins requires a precision that can be met only by enzymes. However, enzymes themselves are proteins. A solution to this problem was perceived after it was shown first that DNA is the genetic material of the cell

(4–6), and secondly that the bihelical structure of DNA and the restrictions of base pairing (7) suggested a semi-conservative mechanism for its replication (8). The possible correlation of the nucleotide sequence of DNA with the amino acid sequence of polypeptides was developed to form the basis of the 'sequence hypothesis'.

4. The sequence hypothesis

This hypothesis has been proven to such an extent that it is scarcely ever mentioned. Three ideas were important for its development; first, that genes controlled the structure of enzymes, epitomized by the 'one gene – one enzyme' concept (9); secondly, that nucleic acids (both DNA and RNA) carry genetic information (4–6,10,11); thirdly, that both polypeptide chains (neglecting folding) and nucleic acids are linear polymers. These considerations led to the notion that the linear nucleotide sequence of the gene codes for the precise linear sequence of amino acids. Nowadays it is often easier to deduce the amino acid sequence of the protein from the nucleotide sequence of the gene, using the genetic code discussed in Section 8 (see *Table 1.3*). In formulating the coding

Table 1.3.   The standard genetic code and non-standard codons used by mitochondria of different species

First base	Second base U	C	A	G	Third base
U	Phe	Ser	Tyr	Cys	U
	Phe	Ser	Tyr	Cys	C
	Leu	Ser	Stop	Stop Trp (m,y,i)	A
	Leu	Ser	Stop	Trp	G
C	Leu Thr(y)	Pro	His	Arg	U
	Leu Thr(y)	Pro	His	Arg	C
	Leu Thr(y)	Pro	Gln	Arg	A
	Leu Thr(y)	Pro	Gln	Arg Trp (p)	G
A	Ile	Thr	Asn	Ser	U
	Ile	Thr	Asn	Ser	C
	Ile	Thr	Lys	Arg Stop (m) Ser (i)	A
	Met (m,y,i*)				
	Met	Thr	Lys	Arg Stop (m) Ser (i)	G
G	Val	Ala	Asp	Gly	U
	Val	Ala	Asp	Gly	C
	Val	Ala	Glu	Gly	A
	Val	Ala	Glu	Gly	G

Unless otherwise indicated the code refers to that established in *E.coli*, which applies also to cytosolic protein synthesis in eukaryotes. Codons found in mitochondria are identified by the following abbreviations: m, mammals; y, yeast; i, insects *(Drosophila melanogaster)*; p, plants (maize). *The sequence AUAA serves as an initiation codon in *Drosophila melanogaster*.

problem a clear distinction was drawn between the 20 common amino acids and the amino acids which were found in only a few proteins and could be derived by secondary modifications after the polypeptide chain had been synthesized. A further simplification was made by recognizing that thymine (5-methyluracil) present in DNA and uracil present in RNA may be regarded as equivalent since both pair specifically with adenine. Similarly, cytosine and its derivative 5-methylcytosine, each of which forms base pairs with guanine, were also regarded as equivalent. The scheme for base pairing, which was first proposed by Watson and Crick (7), has stood the test of time, and their bihelical model of DNA structure has been refined as the result of improved X-ray diffraction data obtained from crystals of appropriate model compounds (12).

The structure of an oligoribonucleotide comprising the four common nucleotides is presented in *Figure 1.3a*. The polyribonucleotide chain can form antiparallel bihelical structures stabilized by $A \cdot U$ and $G \cdot C$ base pairs (see *Figure 1.3*), 0.305 ± 0.0005 nm apart, there being $10-11$ base pairs per turn of the helix (13). The dimensions of the RNA bihelix and of the A form of the DNA bihelix are very similar.

Many RNA species such as transfer RNA (see Chapter 2), ribosomal RNA (see Chapter 3), and messenger RNA (see Chapter 4), which are implicated in protein biosynthesis, form unique structures in which a prominent feature is the folding of the polynucleotide chain to form short bihelical regions (14). These regions, which are stabilized by intramolecular base pair interactions, have the same structural features as the RNA bihelix.

5. The central dogma

The central dogma (1) was put forward to express general ideas concerning the flow of sequence information from one polymer to another. The advantages of a template in the determination of sequence were regarded as paramount. In its simplest form the 'dogma' states that 'once sequence information has passed into protein it cannot get out again'. In a later re-statement of the dogma (15) Crick distinguished between general transfers of sequence information which can occur in all cells; special transfers of sequence information which do not occur in most cells, but may occur in special circumstances; and three transfers of sequence information (protein to protein, protein to DNA, and protein to RNA) which the central dogma postulates never occur. General transfer of sequence information may take place between DNA and DNA, between DNA and RNA, and between RNA and protein (*Figure 1.4*). Special transfers are found in virus-infected cells and these include RNA to DNA in cells infected with RNA tumour viruses, RNA to RNA in cells infected with RNA viruses, and DNA to protein in an artificial cell-free protein synthesis system using single-stranded DNA as template and in the presence of neomycin (*Figure 1.4a*). In each of these cases the transfer of sequence information involves a single-stranded nucleic acid as

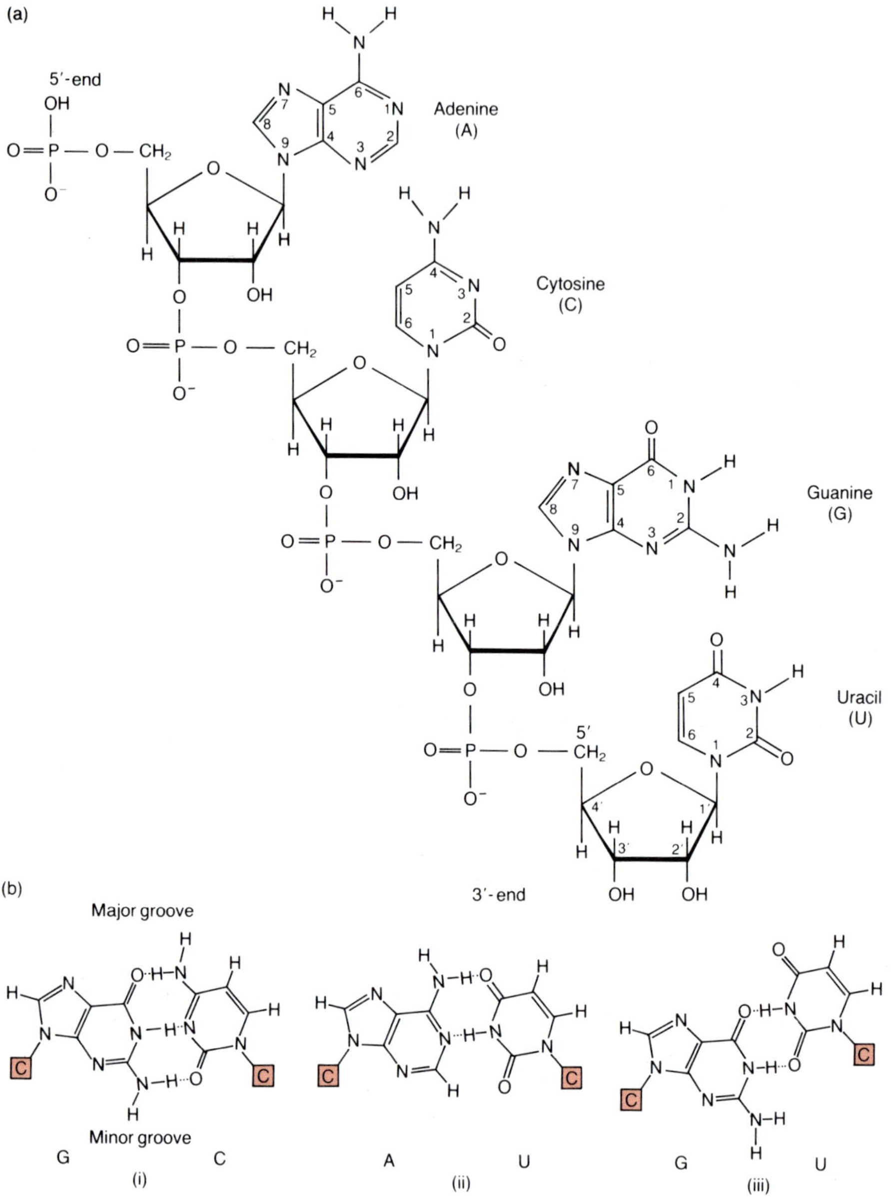

Figure 1.3. Structural elements of ribonucleic acids. (a) Primary structure indicating the numbering system for purines and pyrimidines. Note that an older system for numbering pyrimidines was once used and occasionally may still be encountered. (b) Base pairing interactions commonly found in RNA. (i) G·C base pair; (ii) A·U base pair; (iii) G·U base pair. Pairs i and ii are Watson and Crick base pairs; iii is a special kind of base pairing found in intramolecular bihelical regions. The G is shifted towards the minor groove and U is shifted towards the major groove producing less twist in the helix. The minor groove of the helix is on the side of the base pair with the glycosidic bond, of which the carbon atom is boxed (14). C represents carbon 1 of ribose.

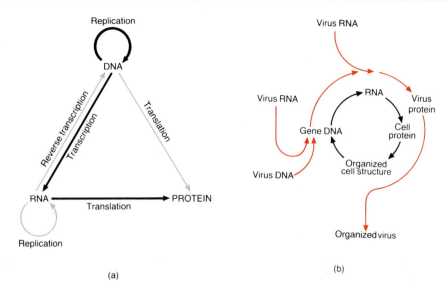

(a)

(b)

Figure 1.4. The flow of genetic information in normal and virus-infected cells. (a) The flow of genetic information according to the central dogma (1,15). Solid arrows show general transfers; dotted arrows show special transfers. The absent arrows are the undetected transfers specified by the central dogma. (b) Illustration of the mechanisms whereby an infecting virus donates either DNA or RNA to the host cell. The viral nucleic acid utilizes the protein synthetic machinery of the host cell and thereby diverts the normal course of cell metabolism in favour of the synthesis of viral components and viral synthesis. Viral RNA may be single-stranded and act directly as mRNA (plus strand), or act as a template for mRNA synthesis (minus strand). Viral RNA may also be double-stranded and one strand is transcribed to produce mRNA. Viral RNA (retrovirus) may also be a template for DNA synthesis using reverse transcriptase. Viral DNA may be either single-stranded or double-stranded. The donation of genetic information from the virus to the cell is in accord with the central dogma summarized in (a).

a template using the simple code adenine → thymine or uracil; guanine → cytosine or 5-methylcytosine; thymine or uracil → adenine; and cytosine or 5-methylcytosine → guanine.

6. The flow of sequence information in protein biosynthesis

The recognition that DNA carries genetic information was followed by the demonstration that DNA replication is semi-conservative, proving that sequence information may pass from DNA to DNA (8). In eukaryotic cells DNA is located in the nucleus, and protein biosynthesis was observed to take place in the cytoplasm on or within microsomes, which are now called ribosomes (see Chapter 3). The absence of DNA from the cytoplasm and the presence of RNA suggested that the template needed for protein biosynthesis was RNA, now called messenger RNA (mRNA). There was little progress until the development of cell-free systems for protein synthesis. The first clear proof of mRNA function was the biosynthesis of polyphenylalanine under the direction of polyuridylic acid (16). Subsequently, Khorana and his colleagues (17) proved beyond any doubt by chemical synthesis that, in protein biosynthesis, the flow of sequence information is from DNA to RNA to protein. Viruses in contributing their own sequence information hijack the translation machinery of the cell (see *Figure 1.4b*).

The flow of sequence information is governed by the polarity of polypeptides (see *Figure 1.1*) and polynucleotides (see *Figure 1.3*). The α-globin chain of haemoglobin was used as a model to establish that biosynthesis of a polypeptide begins at the amino terminus and proceeds sequentially to the carboxyl terminus (18,19). The use of synthetic oligonucleotides of defined sequence revealed that mRNA is translated beginning at the 5'-end with a start codon, continuing towards the 3'-end, and finishing with a termination codon (17,20,21) (see Section 3.3 of Chapter 4). Conventionally, gene sequences and mRNA sequences are presented in the 5' to 3' direction so that the reader encounters the start codon before the termination codon.

7. Properties of mRNA

The sequence information of a gene is copied (transcribed) into an RNA sequence using the complementary strand of DNA as the template, ribonucleoside triphosphates, and the enzyme DNA-dependent RNA polymerase (22). The primary transcript is a single strand of RNA (precursor or pre-mRNA) which is a faithful copy of the DNA sense strand sequence (with U residues in the RNA in place of the T residues found in the DNA).

In organisms which do not have a nucleus (prokaryotes) pre-mRNA usually undergoes little or no modification so that pre-mRNA and mRNA are very similar, if not identical (see Section 2.1 of Chapter 4). Both pre-mRNA and mRNA comprise the coding region of the gene with additional non-translated sequences located at the 5'- and 3'-ends of the gene sequence. These 5'- and 3'-non-translated sequences are implicated in the control of the translation of the mRNA (see Chapter 7). Gene expression usually involves the co-ordinated transcription of several adjacent genes and translation of mRNA sequences may take place while transcription is still in progress.

In cells with a nucleus (eukaryotes) most of the genetic information is stored in the nucleus. Nuclear genes are more complicated than prokaryotic genes because the coding region is often interrupted by intervening sequences (introns). The primary gene transcript (pre-mRNA) is processed within the nucleus, leading to the formation of mRNA. Maturation of pre-mRNA involves the addition of a 'cap' at the 5'-end, removal of introns and usually the addition of a poly(A) tail at the 3'-end (see Section 2.1 of Chapter 4). As with prokaryotic mRNA the coding region is flanked by 5'- and 3'-non-translated sequences. After the maturation process has been completed mRNA is transported to the cytoplasm (23) where it is translated.

All eukaryotes contain subcellular organelles, mitochondria, whose function is to convert the oxidation of metabolic fuels into chemical energy by oxidative phosphorylation. In addition, plant cells also contain a second class of organelle, chloroplasts, which are required for photosynthesis. Both mitochondria and chloroplasts, which are probably of bacterial origin, have separate, distinct systems, including ribosomes and tRNAs, for synthesizing proteins, coded for by the organelle DNA.

One view is that the primitive eukaryotic (proto-eukaryotic) cells which preceded animals and plants had neither mitochondria nor chloroplasts. Subsequently a stable endosymbiotic relationship was established when formerly free-living aerobic bacteria became engulfed in the cytoplasm of the proto-eukaryote, contributing the benefits of oxidative phosphorylation. The remnants of the bacterial partner are now seen as mitochondria. Traces of their ancestry are seen in their machinery for the synthesis of proteins (24,25). Later, a separate endocytotic event involving a cyanobacterium led to the first plant cell with the engulfed bacteria contributing the capacity for photosynthesis and developing into chloroplasts. These ideas are known as the endosymbiont hypothesis (26–28).

7.1 Modification of the information encoded by the mRNA

In eukaryotic cells the mRNA is not only spliced to remove non-informational intron sequences but can also be spliced in different ways (alternative splicing) to yield different polypeptides (22). In addition the information encoded by the mRNA may be modified by editing, for example in kinetoplasts of protozoa.

The editing process leads to functional mRNA by altering the nucleotide sequence of the primary transcript, for example by the deletion of one or several uridine residues encoded in the gene or by the insertion of uridines not encoded in the genomic sequence (a cryptogene). The editing process is an unusual example of gene splicing. A second transcript (a guide RNA) derived from another gene provides a template for the editing process (29,30). mRNA editing can create initiation codons, extend open reading frames, create termination codons, eliminate internal frame shifts and modify the nucleotide sequence within the 3'-non-translated region and the poly(A) tail. RNA editing can be extensive and in some cases may account for half the mRNA sequence (e.g. cytochrome c oxidase III mRNA and ATPase subunit 6 mRNA, ref. 31).

The mitochondrial DNA of kinetoplast protozoa is known as kinetoplast DNA; it comprises a network of approximately 10 000 minicircles and 50 maxicircles which are interlinked (catenated), forming a highly organized structure (32). Kinetoplast DNA contains genes for proteins normally encoded by the mitochondrial DNA. However, mRNAs for several gene products, for example, cytochrome b, subunit III of cytochrome oxidase, and ATPase subunit 6, are obtained by 'RNA editing' of transcripts of cryptogenes (29–31). Cryptogenes are encoded in maxicircles and guide genes are encoded by both mini- and maxicircle DNA.

8. The genetic code

In protein synthesis the information in the coding sequence of mRNA specifies the amino acid sequence of the growing polypeptide chain according to the rules of the genetic code. Early genetic and biochemical experiments established that the code is triplet, comma-less, and non-overlapping. Thus, mRNA is decoded as self-contained groups of three bases (codons) with no punctuation—such as an extra nucleotide to separate codons—nor any bases shared between consecutive codons. This means that a polynucleotide can in principle be decoded in any one of three possible series of triplets (reading frames) depending on the starting point, but this ambiguity is resolved by the binding of mRNA to the small ribosomal subunit and the existence of a special initiator tRNA (see Chapter 4) which ensures that translation starts at a unique codon, usually AUG, within the context of a particular nucleotide sequence. In this way, the reading frame is determined by the initiation codon and the subsequent nucleotide sequence is decoded as consecutive groups of three bases in phase with the first codon. Occasionally, mRNA may contain more than one initiation codon and in such cases more than one polypeptide with different amino acid sequences may be produced by decoding the same nucleotide sequence in phase with each of the initiation codons in separate ribosome cycles.

For many years the standard genetic code, as given in *Table 1.3*, was thought to be universal, each triplet specifying the same amino acid in all organisms. Of the 64 possible triplets, 61 are known to code for the 20 genetically-encoded amino acids found in all, or nearly all, proteins. Since the same amino acid may be specified by up to six different triplets of bases, termed synonymous codons, the genetic code is degenerate. Each codon, however, specifies only one amino acid and there is no ambiguity. Three triplets specify chain termination by a process which does not involve tRNA. Thus all 64 codons are in fact utilized, but only 61 specify the incorporation of amino acids into polypeptides. Fewer than 61 tRNAs are, however, sufficient because the interaction between the first base of the anticodon and the third base of the codon has less stringent requirements than standard Watson–Crick base pairing. This 'wobble hypothesis', originally proposed by Crick (33), has been verified in many cases, as summarized in *Table 1.4*, although some of the predicted base pairs have not been found. As discussed in Chapter 2, modified nucleotides in tRNA contribute

Table 1.4. Base pairing in the wobble position of codons and anti-codons

(a) Cytosolic systems				
First base of anti-codon	G	I	U*	
Third base of codon	U,C	U,C,A	A,G,U	
(b) Mitochondrial systems				
First base of anticodon	C	G	U	U*
Third base of codon	A,G	U,C	U,C,A,G	A,G

*Modified base. I, inosine, is the nucleoside of hypoxanthine (6-hydroxypurine).

to the specificity of codon – anticodon interactions and thus to the accuracy of the decoding process.

8.1 Deviations from the standard genetic code

The first deviation from the above code was discovered by Barrell *et al.* (34) when AUA was found to code for methionine in human mitochondria. It is now known that not only mitochondria but also chloroplasts, as well as certain organisms such as mycoplasma and ciliated protozoa, use codes that differ in certain minor respects from the standard code, as summarized in *Table 1.3*. Variations in the genetic code of different organelles and organisms are thought to have arisen as a result of the loss of some tRNA genes and mutational pressures on DNA giving rise to predominance of either AT- or GC-rich codons (35).

Another deviation from the standard code accounts for the synthesis of selenoproteins by a novel mechanism operating in both eukaryotes and prokaryotes. Mammalian glutathione peroxidase and bacterial formate dehydrogenase contain a selenocysteine residue at the active site and this amino acid residue is encoded by UGA, which usually functions as a termination codon (36). It has been reported that a novel tRNA species, which is aminoacylated by serine and has an anticodon complementary to UGA, is involved in the incorporation of selenocysteine into selenoproteins in *Escherichia coli*, presumably by conversion of the serine residue bound to the tRNA into selenocysteyl tRNA (37). Thus, this tRNA not only acts as a suppressor of chain termination but also functions to insert an unusual amino acid into a nascent protein during polypeptide chain elongation. This finding constitutes the first exception to the long-held view that only the 20 common amino acids are genetically encoded by the DNA.

9. The adaptor hypothesis and bilingual tRNA

Once the mRNA has been transcribed and, in eukaryotes, transported to ribosomes within the cytoplasm, the genetic information has to be translated from a nucleotide sequence into an amino acid sequence. The adaptor hypothesis proposed by Crick in 1958 (1) states:

RNA presents mainly a sequence of sites where hydrogen bonding could occur. One would expect, therefore, that whatever went on to the template in a specific way did so by forming hydrogen bonds. It is therefore a natural hypothesis that the amino acid is carried to the template by an 'adaptor' molecule, and that the adaptor is the part which actually fits on to the RNA. In its simplest form one would require twenty adaptors, one for each amino acid.

What sort of molecules such adaptors might be is anyone's guess. But there is one possibility which seems inherently more likely than any other – that they might contain nucleotides. This would enable them to join on to the RNA template by the same 'pairing' of bases as is found in DNA, or in polynucleotides.

Soluble RNA (now termed transfer RNA or tRNA) with amino acids attached was discovered by Hoagland, Zamecnik and Stephenson (38). It soon became apparent that these were the postulated adaptors which interacted with mRNA through base pairing of the anti-codon within its sequence. Thus aminoacyl-tRNA is bilingual, linking a particular codon to a particular amino acid to allow the nucleotide sequence of the gene to be translated into the amino acid sequence of the protein. The structure and function of tRNA are discussed in Chapter 2.

10. Origin of multimeric proteins and translocation of polypeptides

Proteins consisting of two or more subunits can originate by two alternative routes. One pathway involves the synthesis of a single-chain polypeptide precursor which is processed after release from the ribosome by proteolytic cleavage to produce the subunits of the final protein. An example of this process is the biosynthesis of insulin from proinsulin by excision of the connecting peptide which in the precursor links the A and B chains. A variant of the synthesis of subunits of multimeric proteins from a common precursor is the synthesis of several individual polypeptides by proteolytic cleavage of a so-called polyprotein. This process is characteristic of the synthesis of some hormonally-active peptides such as adrenocorticotrophic hormone (ACTH), melanotrophins, and β-endorphin. The second route involves the transcription of separate genes into the corresponding mRNAs and their independent translation into polypeptides which then associate spontaneously to form a multimeric protein. In the case of haemoglobin, the α- and β-chains are synthesized in approximately equal amounts by control of the amounts of the mRNAs and their translation rates. There are two genes for α-globin but only one for β-globin and as a consequence transcription gives rise to an excess of α-globin messenger. Usually the ratio of α-/β-globin mRNA is approximately 1.7, but translation of α-globin mRNA is less efficient and there is only an approximately 10% overproduction of α-globin. After insertion of haem and assembly of the haemoglobin tetramer, excess free α-chains are degraded thus limiting the accumulation of free α-chains and preventing their precipitation. Although cytosolic proteins such as haemoglobin are retained in the cytoplasm, many proteins are transported from their site of synthesis to remote locations; the mechanisms involved in this translocation process are considered in Chapter 6.

11. The translational machinery

The involvement of RNA in protein biosynthesis was deduced from the finding of a direct relationship between the amount of RNA in a cell and the rate of protein biosynthesis. It was also established that RNA is localized mainly in the cytoplasm. Early experiments in subcellular fractionation led to the isolation of microsomes which were always found to contain RNA. Subsequently microsomes were identified *in situ* by electron microscopy. The term ribosome was coined by

Roberts (39) for these small electron-dense particles of $2.7 \times 10^6 - 4 \times 10^6$ Da comprising $40 - 65\%$ RNA and little or no lipid.

The confirmation of the adaptor hypothesis not only stimulated the belief in a genetic code and hastened its elucidation, but also focused attention on the ribosome as the site of decoding of the genetic information. Ribosomes constitute a substantial part of a rapidly growing cells, such as bacteria, reflecting their major role in cell metabolism. A single *E.coli* bacterium contains approximately 16 000 ribosomes which account for about 60% of the dry weight of the cell. Eukaryotic cells are larger and their cytoplasm contains more ribosomes; for example somatic cells contain approximately 1×10^6 ribosomes. However, the very large oocytes of the toad, *Xenopus laevis,* may contain as many as 1×10^{12} ribosomes.

An early discovery of major significance was that ribosomes can be readily dissociated into two unequal subunits by manipulating the ionic environment, or more precisely the ratio of monovalent cations to divalent cations such as magnesium ions. For example, in a 50 mM Hepes buffer (pH 7.5) containing 150 mM KCl and 10 mM $MgCl_2$ bacterial ribosomes sediment as 70S particles. However, when the $MgCl_2$ concentration is reduced to 1 mM, two subunits or subparticles (smaller and larger) are produced. The association of ribosomal subunits to form ribosomes and the dissociation of ribosomes into subunits take place under physiological conditions during ribosome function and are principal events in the ribosome cycle (see Chapter 4). The ribosome is essential for the correct recognition and translation of the genetic message encoded in mRNA. The peptide bond is formed on the larger subunit, with the smaller subunit playing a leading part in binding the mRNA. The larger subunit has at least two and probably three binding sites for tRNA derivatives, the A-, P-, and E-sites (40). The aminoacyl-tRNA carrying the incoming amino acid is bound at the A-site, with peptidyl-tRNA being already in place at the P-site. The E-site is occupied by the exiting deacylated tRNA before it leaves the ribosome. The peptidyl transferase activity of the larger subunit is manifest as follows: the nascent polypeptide chain is transferred to aminoacyl-tRNA which now becomes peptidyl-tRNA, and the uncharged tRNA is released back into the tRNA pool. However, protein synthesis cannot continue while peptidyl-tRNA is bound at the A-site. The peptidyl-tRNA is returned to the P-site by a translocation step leaving the A-site vacant for another aminoacyl-tRNA as specified by the next codon. Each translocation step leads to the relative movement of mRNA and the ribosome by one codon.

The time taken for the synthesis of a polypeptide chain of about 140 amino acids (e.g. one of the haemoglobin chains) is about 20 sec at 37°C. Different rates might be expected in other systems but the rate quoted is likely to be of the right order. The rate increases on raising the temperature and decreases on cooling (41).

Thus, in order to decode the genetic message and carry out sequential peptide bond formation, the ribosome is required to accomodate at its active site a complex of mRNA, aminoacyl tRNA, peptidyl tRNA and appropriate protein factors with energy supplied by ATP and GTP. Factors are needed for all the different stages of the initiation of polypeptide synthesis, the elongation of the nascent polypeptide chain, the termination of protein synthesis with the release of the completed polypeptide chain and of mRNA, and the translocation step in which mRNA and ribosome move relative to one another by three nucleotides in order to bring the next codon into the reading frame. The well-being of the cell requires that each step in protein biosynthesis is carried out with high fidelity (42).

Except for a few small polypeptides such as gramicidin-S and glutathione, all proteins—including enzymes, immunoglobulins, blood pigments such as haemoglobin, structural proteins such as collagen, muscle proteins such as actin and myosin, photosynthetic proteins such as rhodopsin, certain hormones of 20–25 amino acids in size, such as ACTH and growth hormone, tripeptides such as thyrotropic hormone, and ribosomal proteins themselves—are synthesized by the same basic mechanism discussed above, which appears to be general throughout the living world.

Both gene structure and transcriptional control of gene expression are very important aspects of protein biosynthesis covered in detail in the companion book on gene structure and transcription (22). Hence, the following chapters will focus on the mechanisms and control of the process by which the nucleotide sequence of mRNA is translated into an amino acid sequence, and on the structure of the components needed for translation.

12. Further reading

Nobel lectures in molecular biology, 1933 – 1975. (1977) Elsevier, New York.

Cantor,C.R. and Schimmel,P.R. (1980) *Biophysical chemistry. Part I: the conformation of biological macromolecules.* W.H.Freeman and Co., San Francisco.

Richardson,J.S. (1981) The anatomy and taxonomy of protein structure. *Adv. Protein Chem.,* **34**, 167 – 339.

Saenger,W. (1984) *Principles of nucleic acid structure.* Springer-Verlag, Berlin.

Prog. Nucleic Acid Res. Mol. Biol., Vol. 19, Symposium on mRNA: the relation of structure and function (1976).

Woese,C.R. (1967) The present status of genetic code. *Prog. Nucleic Acid Res. Mol. Biol.,* **7**, 107 – 72.

13. References

1. Crick,F.H.C. (1958) *Symp. Soc. Exp. Biol.,* **12**, 138 – 63.
2. Kannan,K.K., Liljas,A., Waara,I., Bergsten,S., Lougren,S., Strandberg,B., Bengtsson,U., Carlbom,U., Fridborg,K., Jarup,L., and Petef,M. (1972) *Cold Spring Harbor Symp. Quant. Biol.,* **36**, 221 – 31.
3. Darnall,D.W. and Klotz,M. (1975) *Arch. Biochem. Biophys.,* **166**, 651 – 82.
4. Griffith,F. (1928) *J. Hyg.,* **27**, 113 – 59.
5. Avery,O.T., McLeod,C.M., and McCarty,M. (1944) *J. Exp. Med.,* **79**, 137 – 57.
6. Hershey,A.D. and Chase,M. (1952) *J. Gen. Physiol.,* **36**, 39 – 56.
7. Watson,J.D. and Crick,F.H.C. (1953) *Nature,* **171**, 964 – 7.
8. Meselson,M. and Stahl,F.W. (1958) *Proc. Natl. Acad. Sci. USA,* **44**, 671 – 82.

9. Beadle,G.W. and Tatum,E.L. (1941) *Proc. Natl. Acad. Sci. USA*, **27**, 499–506.
10. Gierer,A. and Schramm,G. (1956) *Nature*, **177**, 702–3.
11. Fraenkel-Conrat,H. (1956) *J. Am. Chem. Soc.*, **78**, 882.
12. Drew,H.R., McCall,M.J., and Calladine,C.R. (1988) *Annu. Rev. Cell Biol.*, **4**, 1–20.
13. Neidle,S. (ed.) (1981) *Topics in nucleic acid structure*. MacMillan, London.
14. Wyatt,J.R., Puglisi,J.D., and Tinoco,I. (1989) *BioEssays*, **11**, 100–6.
15. Crick,F.H.C. (1970) *Nature*, **227**, 561–3.
16. Nirenberg,M.W. and Mattaei,J.H. (1961) *Proc. Natl. Acad. Sci. USA*, **47**, 1588–602.
17. Kossel,H., Morgan,A.R., and Khorana,H.G. (1967) *J. Mol. Biol.*, **26**, 449–75.
18. Bishop,J., Leahy,J., and Schweet,R. (1960) *Proc. Natl. Acad. Sci. USA*, **46**, 1030–8.
19. Dintzis,H.M. (1960) *Proc. Natl. Acad. Sci. USA*, **47**, 247–61.
20. Salas,M.A., Smith,M.A., Stanley,W.M., Wahba,A.J., and Ochoa,S. (1965) *J. Biol. Chem.*, **240**, 3988–95.
21. Thach,R., Cecere,M.A., Sundararajan,T.A., and Doty,P. (1965) *Proc. Natl. Acad. Sci. USA*, **54**, 1167–73.
22. Beebee,T. and Burke,J. (eds) (1988) *Gene structure and transcription*. IRL Press, Oxford.
23. Green,M.R. (1990) *Current Opinion Cell Biol.*, **1**, 519–25.
24. Gray,M.W. (1988) *Biochem. Cell. Biol.*, **66**, 325–48.
25. Van de Peer,Y., Neefs,J.-M., and De Wachter,R. (1990) *J. Mol. Evol.*, **30**, 463–76.
26. Schwartz,R.M. and Dayhoff,M.O. (1978) *Science*, **199**, 395–403.
27. Doolittle,W.F. (1980) *Trends Biochem. Sci.*, **9**, 146–9.
28. Gray,M.W. (1989) *Annu. Rev. Cell Biol.*,**5**, 25–50.
29. Blum,B., Bakatara,N., and Simpson,L. (1990) *Cell*, **60**, 189–96.
30. Simpson,L. (1990) *Science*, **250**, 512–13.
31. Bhat,G.J., Koslowsky,D.J., Feagin,J.E., Smiley,B.I., and Stuart,K. (1990) *Cell*, **61**, 885–94.
32. Simpson,L. (1986) *Int. Rev. Cytol.*, **99**, 119–79.
33. Crick,F.H.C. (1966) *J. Mol. Biol.*, **19**, 548–55.
34. Barrell,B.G., Bankier,A.T., and Druin,J. (1979) *Nature*, **282**, 189–94.
35. Osawa,S. and Jukes,T.H. (1988) *Trends Genet.*, **4**, 191–8.
36. Engelberg-Kulka,H. and Schoulaker-Schwarz,R. (1988) *Trends Biochem. Sci.*, **13**, 419–21.
37. Leinfelder,H., Zehelein,E., Mandrand-Bertholet,M.-A., and Bock,A. (1988) *Nature*, **331**, 723–5.
38. Hoagland,M.B., Zamecnik,P.C., and Stephenson,M.L. (1957) *Biochim. Biophys. Acta*, **24**, 215–16.
39. Roberts,R.B. (1958) In Roberts,R.B. (ed.) *'Microsomal particles and protein biosynthesis'*, Pergamon Press, London; p. 9.
40. Nierhaus,K.H. (1990) *Biochemistry*, **29**, 4997–5008.
41. Cox,R.A., Pratt,H., Huvos,P., Higginson,B., and Hirst,W. (1973) *Biochem. J.*, **134**, 775–93.
42. Parker,J. (1989) *Microbiol. Rev.*, **53**, 273–98.

2

Charging of tRNA by aminoacyl-tRNA synthetases

1. Introduction

Aminoacyl-tRNAs are key intermediates in the incorporation of amino acids into polypeptide chains and their synthesis performs a dual function. First, aminoacylation of tRNA requires ATP, which provides the energy for peptide bond synthesis. Secondly, the process allows decoding of the genetic information by antiparallel base pairing between the three bases of mRNA codons and the complementary anticodons of tRNA during peptide bond formation on the ribosome. This interaction involves conventional (Watson – Crick) base pairs between the first and middle base of the codon and the third and middle base of the anticodon, respectively. The third base of the codon, however, pairs with the first base of the anticodon (from the 5'-terminus) by a less stringent interaction (see *Table 1.4*). This so-called 'wobble' reduces the tRNA species required to decode the 61 sense codons of the genetic code to fewer than the theoretical number. Both prokaryotes and the cytosol of eukaryotes contain only a few more than 40 different tRNA species, whereas mitochondria and chloroplasts contain an even smaller number of tRNAs which differ in structure from the cytosolic species. Thus, in animal and yeast mitochondria there are only 22 and 24 tRNAs, respectively, as deduced from the number of tRNA genes in mitochondrial DNA. This small number suffices for the mitochondrial translation system because of deviations from the standard genetic code and even less stringent requirements for wobble interactions, which allow up to four bases in the 3'-position of mRNA codons to be read by the same base in the 5'-position of the anticodon.

2. Structure of tRNA

The nucleotide sequences of more than 350 tRNAs from different organisms have been established (1). All tRNAs are single-stranded molecules about 80 nucleotides in length and the primary sequence may be arranged so as to maximize base pairing, giving a secondary structure resembling a clover leaf shape, consisting of four stem regions and loops, including one loop of variable size containing 5 – 21 nucleotides (*Figure 2.1a*). The tertiary structures of several tRNAs, which have been elucidated by X-ray crystallography, show a compact L-shaped molecule which is stabilized by hydrogen bonding between widely separated bases, particularly in the loops of the clover leaf, and hydrophobic stacking interactions, as illustrated in *Figure 2.2* (4).

The acceptor stem is formed through base pairing between the 5′- and 3′-terminal regions of tRNA. The – CCA 3′-sequence, which is not base paired and is common to all tRNAs, is added to the 3′-terminus as one of the post-transcriptional modifications in tRNA biosynthesis. The 3′-terminal A residue of the tRNAs accepts the amino acid in the aminoacyl-tRNA synthetase reaction. Thus, the amino acid is added at a site which is distal to the anticodon in the tertiary structure. The invariant – CCA sequence is also important for the recognition of charged tRNAs by the prokaryotic elongation factor EF-Tu and GTP, as well as for the subsequent transfer of the aminoacyl-tRNA to the ribosomal A-site (2). Replacement of the amino acid residue by a hydroxy acid analogue lowers the binding efficiency of EF-Tu by a factor of 300 and uncharged tRNA or tRNA acylated with a D-amino acid residue also interact only weakly. Modification of the – CCA sequence, for example by insertion of an additional C residue or replacement of the penultimate C by U, weakens the interaction with EF-Tu and it is thought, therefore, that the elongation factor recognizes the overall spatial structure of the stacked – CCA end as well as the 3′-terminal A and the – NH_3^+ of the aminoacyl group (2).

A characteristic structural feature of tRNAs from all organisms is the presence of many different modified nucleosides. The modifications range from the simple methylation of bases or ribose sugar residues to very complex substitutions (*Figures 2.1* and *2.3*; *Table 2.1*). All modifications take place after transcription of the tRNA genes, usually by alteration rather than replacement of bases. Occasionally, however, base exchange occurs as, for example, in the insertion of inosine or the unusual base queuosine which involves replacement of adenine or guanine in the precursor polynucleotide. The importance of modified nucleosides in tRNA is thought to be related to the stability of the tertiary structure and to contribute to codon – anticodon interactions, prevention of mischarging of tRNA by inappropriate amino acids (3), increased translational efficiency and fidelity, and maintenance of the reading frame. In studies of mutants lacking tRNA modifying enzymes it has been found that changes in the modified nucleoside adjacent to the 3′-side of the anticodon (position 37) or in the wobble position of the anticodon itself (position 34) are particularly important, but modified nucleosides in other positions, for example pseudouridine in positions 38, 39, and 40 of eubacterial tRNA, also appear to influence the elongation rate and error level in translation.

(a)

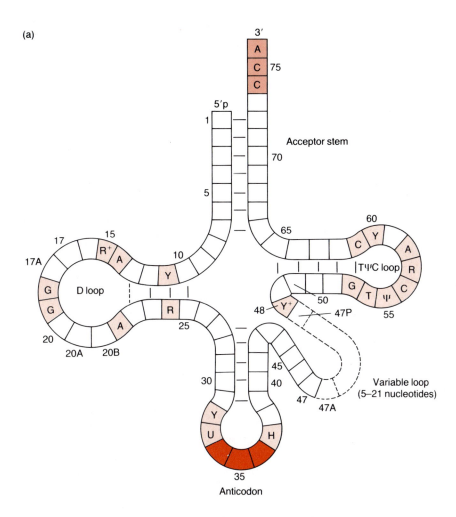

Figure 2.1. Structure of transfer RNA. **(a)** Generalized secondary structure showing the positions of invariant and semi-invariant bases in all tRNA sequences except for initiator tRNAs. The numbering system is that used for yeast tRNA^Phe with additional bases being assigned letters. Y = pyrimidine, R = purine, H = hypermodified purine. R^+ and Y^+ are usually complementary, positions 9 and 26 are usually purines and position 10 is usually G or a modified G. The variable loop ranges in size from 5 to 21 nucleotides and the D-loop also shows some variation in size, which is, however, relatively minor.

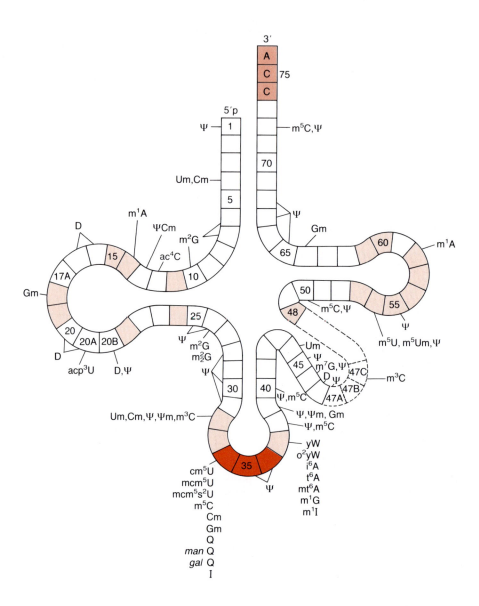

Figure 2.1. Structure of transfer RNA. **(b)** Structure of eukaryotic transfer RNA showing the variety and distribution of modified nucleosides in different tRNAs. The most common modified nucleoside is pseudouridine (ψ) and other common modifications include methylation of either bases or the 2-position of ribose. Abbreviations are listed in *Table 2.1.*

Table 2.1 Modified nucleosides in eukaryotic tRNA

Name of modified nucleoside	Abbreviation	Positions in tRNAs
Pseudouridine	ψ	1,13,20B,25,26,28,30,31, 32,35,36,38,40,45,46, 47A,50,54,55,65,67,68,72
2'-O-methyl nucleosides	Um,Cm,Gm, ψm	4,13,18,32,34,39,44,64
N^2-methylguanosine	m^2G	6,7,10,26
N^4-acetylcytidine	ac^4C	12
N^1-methyladenosine	m^1A	14,58
5,6-dihydrouridine	D	16,17,20,20A,20B,47
3-(3-amino-3-carboxy-propyl)uridine	acp^3U	20A
N^2,N^2-dimethylguanosine	m_2^2G	26
3-methylcytidine	m^3C	32,47C
5-carbamoylmethyluridine	cm^5U	34
5-methoxycarbonylmethyluridine	mcm^5U	34
5-methoxycarbonylmethyl-2-thiouridine	mcm^5s^2U	34
5-methylcytidine	m^5C	34,40,50,72
Queuosine*	Q	34
β-D-mannosylqueuosine	*man*Q	34
β-D-galactosylqueuosine	*gal*Q	34
Inosine	I	34
Wybutosine[#] (formerly Y)	yW	37
Wybutoxosine[#] (peroxywybutosine)	o^2yW	37
N^6-isopentenyladenosine	i^6A	37
N^6-threoninocarbonyladenosine	t^6A	37
N^6-methylt^6A	mt^6A	37
N^1-methylguanosine	m^1G	37
N^1-methylinosine	m^1I	37
2'-O-methylpseudouridine	ψm	38
N^7-methylguanosine	m^7G	46
5-methyluridine (ribothymidine)	m^5U	54
2'-O-methyl-5-methyluridine	m^5Um	54

The list gives the names and abbreviations of modified nucleosides found in different eukaryotic tRNAs and the positions where they occur, as shown in *Figure 1b*.

*Q derivatives; [#]Wye derivatives (formerly Y derivatives).

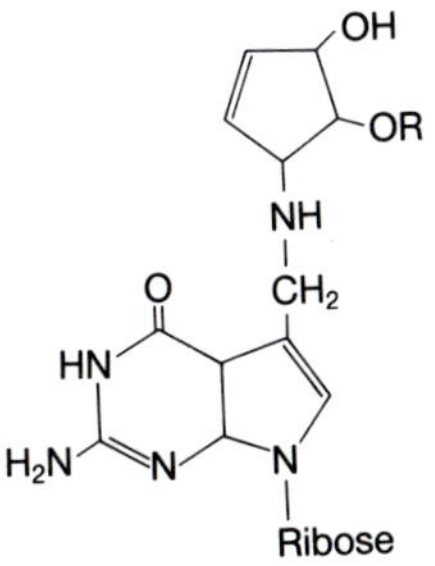

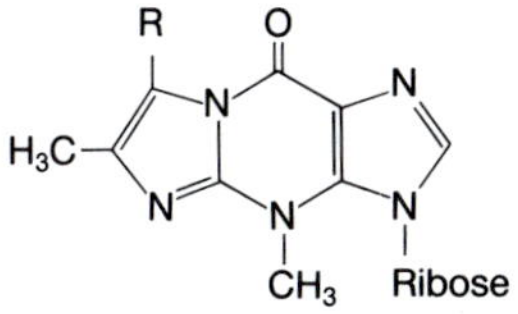

Q or Quo: R = H
(Queuosine)
man Q: R = β-D-Mannosyl
gal Q: R = β-D-Galactosyl

Wyo = Wyosine (formerly 'Yt') R = H

Wybutosine (formerly 'Y')
R = · CH_2 · CH_2 · CH · NH · $COOCH_3$
|
$COOCH_3$

Peroxywybutosine (formerly 'Yr, Yw or peroxy Y')
R = · CH_2 · CH · CH · NH · $COOCH_3$
| |
OOH $COOCH_3$

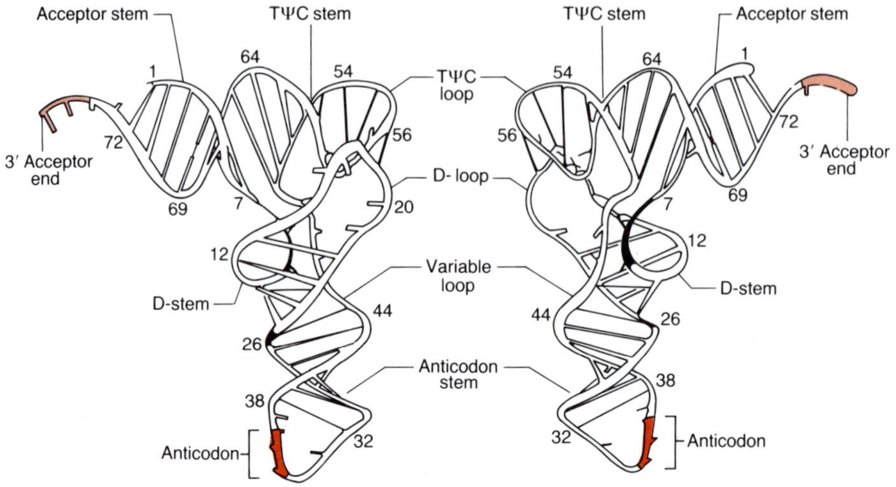

Figure 2.2. Tertiary structure of transfer RNA (4). Two side views of yeast tRNA[Phe] are shown as a schematic diagram of the ribose–phosphate backbone, the numbers referring to nucleotide residues in the sequence. Different parts of the molecule are distinguished by different shading. Hydrogen bonding between bases is shown as cross-rungs and tertiary interactions between bases are indicated by solid black rungs which represent one, two or three hydrogen bonds. Bases not involved in hydrogen bonding are shown as shortened rods attached to the polynucleotide backbone.

Mitochondria and chloroplasts contain all the tRNAs required for protein synthesis by the organelles. The mitochondrial tRNAs contain 59–75 nucleotides and are thus considerably smaller than the cytoplasmic species. Surprisingly, mammalian mitochondria contain only a single gene encoding both initiator and elongator mitochondrial tRNA[Met]. Apparently, the primary transcript is modified in different ways to give rise to the two mitochondrial tRNA[Met] species whereas usually they are transcribed from separate genes.

3.   The aminoacyl-tRNA synthetases

3.1. Occurrence, properties, and structure

All cells active in protein biosynthesis contain aminoacyl-tRNA synthetases (also called amino acid tRNA ligases). A large number of aminoacyl-tRNA synthetases have been isolated from different sources, purified and characterized and the amino acid sequences of some 22 different enzymes have been established. As a general rule, only one synthetase is required for charging isoaccepting tRNAs, even when there are considerable differences in nucleotide sequence as for example in the case of the methionyl-tRNAs specific for polypeptide chain initiation and elongation (*Figure 2.3*). Usually, each enzyme is specific for only one of the 20 genetically encoded amino acids and their cognate tRNAs, but

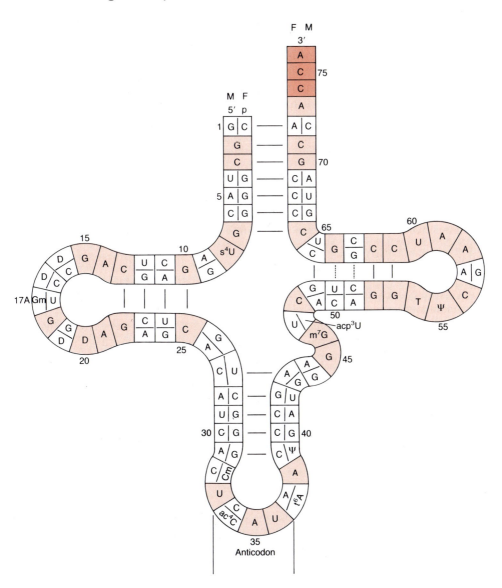

Figure 2.3. Comparison of the nucleotide sequences of the initiator and elongator methionyl-tRNA from *Escherichia coli*. F, initiator tRNA[Met]; M, elongator tRNA[Met]. Sequence data from Sprinzl *et al.* (1). Abbreviations for modified nucleotides are listed in *Table 2.1*. Nucleotides common to both tRNAs are coloured light orange.

exceptions are known. For example, in *Bacillus megaterium* and *Bacillus subtilis* the glutamyl synthetase charges glutaminyl-tRNA with glutamate, the Glu-tRNA^{Gln} being subsequently converted into Gln-tRNA^{Gln} by a specific amidotransferase using glutamine or aspartate as the nitrogen donor, thus correcting the initial misacylation (5).

Prokaryotes contain only 20 aminoacyl-tRNA synthetases, but in eukaryotes, different enzymes are present in the cytosol, mitochondria, and chloroplasts. Sequence studies have shown that in some cases, for example yeast threonyl-tRNA synthetase, cytosolic and mitochondrial enzymes are encoded by separate nuclear genes, but in others such as histidyl-tRNA synthetase both enzymes are encoded by a single nuclear gene (6). The two histidyl-tRNA synthetases have 520 amino acids in common, but the mitochondrial synthetase has an additional 20 amino acids at the amino terminus, suggesting that this sequence may be involved in transporting the primary translation product into the organelle, whereas its absence causes the enzyme to remain in the cytosol.

Although all synthetases catalyse the same reaction there are marked differences in their quaternary structure and size. Thus, synthetases may consist of monomers (α), homodimers (α_2) or homo- and hetero-tetramers (α_4 or $\alpha_2\beta_2$) with polypeptide chains containing from about 300 to 900 amino acid residues. Some of the size differences result from extensions at the amino terminus which appear to be unrelated to the aminoacylation function. A characteristic feature of cytosolic aminoacyl-tRNA synthetases of higher eukaryotes is the formation of specific aggregates. Thus, a high molecular mass complex ($M_r \approx 10^6$ Da) consisting of 11 polypeptides is active in charging nine different tRNAs with amino acids. Some complexes occur bound to the endoplasmic reticulum, others are also found free in the cytosol (7).

3.2. Recognition of tRNAs by aminoacyl-tRNA synthetases

Aminoacyl-tRNA synthetases have recognition sites for specific amino acids, for their cognate isoaccepting tRNAs, and for ATP. As shown by X-ray crystallography of tyrosyl- and methionyl-tRNA synthetases and sequence homologies in several other enzymes, the ATP binding site and the domains for aminoacyl adenylate synthesis are located in the amino terminal region of the molecule with involvement of a characteristic histidyl·isoleucyl (or leucyl)·glycyl·histidine sequence. Analysis of gene deletions of *E.coli* alanyl-tRNA synthetase has confirmed that the amino terminal region is involved in the synthesis of the aminoacyl adenylate, whereas the binding site for tRNA was found to be located on the carboxyl terminal side of this domain. Deletion of 99 amino acids from the carboxyl terminal region of tyrosyl-tRNA synthetase gives a fragment which is still able to synthesize tyrosyl adenylate but no longer binds or aminoacylates the tRNA (8). Analysis of the structure – activity relationship of tyrosyl-tRNA synthetase by site-directed mutagenesis and protein engineering (9) has further clarified the active site of this enzyme and the nature of the interactions involved (*Figure 2.4*). Data obtained by photochemical crosslinking of the components in synthetase – tRNA complexes or by protection of tRNA

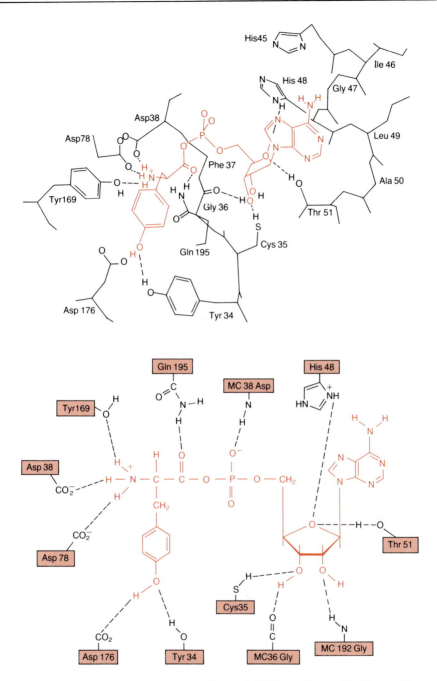

Figure 2.4. Structure of the active site of tyrosyl-tRNA synthetase (9). The top diagram shows the binding site for tyrosyl adenylate (orange) and the bottom sketch indicates the amino acid residues which are probably involved in the binding reaction by forming hydrogen bonds. MC indicates interactions involving peptide bonds of the main chain.

bases by the enzyme against attack by chemical reagents have been used to develop a model (10) which envisages the binding of the synthetase along the inside face of the three-dimensional L-shaped tRNA structure (*Figure 2.2*).

The importance of individual bases and nucleotide sequences for the specific interaction between tRNAs and aminoacyl-tRNA synthetases has been extensively investigated by several different methods, including the chemical modification of tRNAs by specific reagents or UV irradiation, the enzymic dissection of tRNAs into defined fragments and the effect of specific changes in the nucleotide sequence by mutagenesis.

Early experiments led to the conclusion that the 5'-terminal phosphate and the common −CCA sequence at the 3'-terminus are not important for the binding of tRNAs to the synthetase. The anticodon is generally also not significant for recognition by the enzyme since isoaccepting tRNAs, which have different anticodons, are charged by the same synthetase. Moreover, fragments of *E.coli* phenylalanyl-tRNA obtained by removing large portions of the anticodon and D-loops retain the ability to accept the amino acid. In other tRNAs, however, changes in the anticodon, either by chemical modification or removal of a base, resulted in loss of ability to be aminoacylated. Recognition of modified tRNAs by the synthetase was measured either directly by the ability of the synthetase to accept the correct amino acid or indirectly by competition with the original tRNA in the charging reaction. Measurements of the aminoacylation of a modified tRNA are unambiguous since the reaction requires correct interaction of the tRNA with the synthetase, but the interpretation of competition experiments is less certain because the aminoacylation reaction may be inhibited by the modified tRNA or tRNA fragments.

Sometimes even a single base change can alter the specificity of the tRNA−synthetase interaction. Thus, a change from an A to a G residue in position 82 of the SU_{III}^+ tyrosine suppressor tRNA gives a tRNA which accepts glutamine as well as tyrosine. Detailed analysis of a series of mutants of *E.coli* amber suppressor alanyl-tRNA has shown that the $G3 \cdot U70$ base pair is the major determinant for recognition by the alanyl-tRNA synthetase *in vivo* and *in vitro* (*Figure 2.5*). Changes of this base pair to either $G \cdot C$ or $A \cdot U$ prevented aminoacylation with alanine, whereas introduction of this base pair into the corresponding positions of cysteine or phenylalanine amber suppressor tRNAs enables these tRNAs to accept alanine in the aminoacylation reaction even though in other positions there are 38 and 31 differences, respectively, between these two RNAs and alanyl-tRNA (11). Since the same $G3 \cdot U70$ base pair is present also in two *E.coli* isoaccepting tRNAs, $tRNA^{Ala/GGC}$ and $tRNA^{Ala/UGC}$, and is absent from all other tRNAs, it seems to be sufficient in principle for recognition of the tRNA by the synthetase. However, the $G3 \cdot U70$ $tRNA^{Phe/CUA}$ also can be charged to some extent by phenylalanine and hence other determinants evidently allow recognition by the phenylalanyl-tRNA synthetase. Moreover, the efficiency of aminoacylation of the $G3 \cdot U70$ $tRNA^{Cys/CUA}$ by alanine is low. The importance of the $G3 \cdot U70$ base pair in alanyl-tRNA for recognition by the synthetase has also been demonstrated in aminoacylation experiments using a

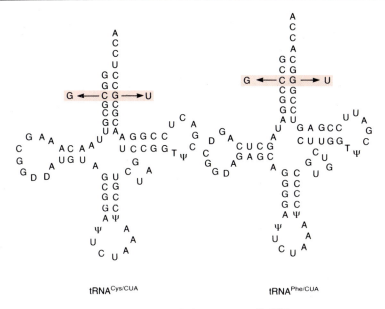

Figure 2.5. Sequences of tRNA^Cys/CUA and tRNA^Phe/CUA (11). The position of the G3·U70 base pair which confers alanine acceptance on these tRNAs is indicated by arrows.

synthetic 'minihelix' composed of only 36 nucleotides representing the amino acid acceptor and TΨC helices, that is nucleotide residues 1–8 and 48–76, of alanyl-tRNA (*Figure 2.6*). The maximal velocity of aminoacylation of this minihelix is similar to that of the tRNA but the K_M is more than five times higher, suggesting less efficient binding to the enzyme. An even smaller 'microhelix' based only on the nucleotide sequence of the acceptor stem of tRNA^Ala also can be aminoacylated by the synthetase but at a much reduced rate (12). These experiments show that the minimum requirement for specific aminoacylation of some tRNAs involves the recognition of only a single base-pair in the acceptor stem, although other nucleotides are also involved in interactions with the synthetase, as shown by protection against nuclease digestion (12). These regions of tRNA may be important in relation to editing and the kinetics of the aminoacylation reaction.

It is also clear that the 3·70 base pair is not universally involved in the recognition of tRNA by synthetases because many tRNAs have identical bases (other than G·U) in these positions. Moreover, in the case of yeast tRNA^Phe, kinetic analysis of the aminoacylation of mutants has shown that five nucleotides, including four in the anticodon loop and one in the D-loop, are critical for recognition by the synthetase provided that they are properly positioned in the tertiary structure (13), which may depend on additional determinants elsewhere.

To summarize, there are major differences between the various tRNAs with respect to the bases involved in their interaction with synthetases and no general rule has emerged. The structural requirements for the recognition of tRNAs by

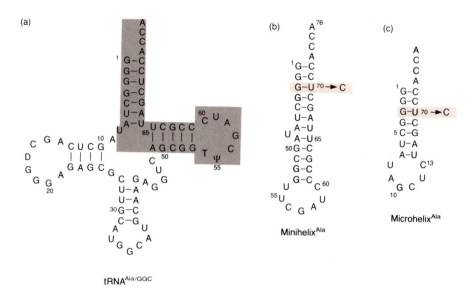

Figure 2.6. Sequence and cloverleaf structure of tRNA$^{Ala/GGC}$ and related alanine-accepting synthetic polyribonucleotides (12). The nucleotide sequence in alanyl-tRNA (**a**) used to construct the minihelix (**b**) is indicated by shading. In a variant of this construct C70 replaced U70. The numbering is based on that used for yeast tRNA^{Ala}. The microhelix (**c**) consisted of nucleotides 1–13 of the acceptor and dihydrouridine arms of the tRNA joined to nucleotides 66–76 of the acceptor stem.

synthetases vary for different species and seem to include both positive and negative determinants in the nucleotide sequence of the tRNA. The former are essential for the correct charging of the tRNA with a specific amino acid, whereas the latter prevent mischarging with incorrect amino acids by inappropriate synthetases. The interplay between such positive and negative determinants is illustrated by the codon and amino acid specificities of the minor *E.coli* isoleucine tRNA, which has a novel modified nucleoside, lysidine (L), in the first position of the anticodon LAU (position 34). This modification is derived from a cytidine residue after transcription and is essential for the recognition of the isoleucine codon AUA by the tRNA. Failure to modify this cytidine residue changes the anticodon to CAU, which not only recognizes the methionine codon AUG but also allows the tRNA to be charged with methionine instead of isoleucine. Thus, although isoleucyl-tRNA synthetase has no absolute requirement for a specific base in position 34 and charges efficiently both the major and minor isoleucine tRNA species (which have, respectively, G and L in this position) the presence of the unmodified C results in discrimination by the enzyme against charging the tRNA with isoleucine. As a consequence, correct aminoacylation and codon recognition remained closely associated during evolution of this tRNA base modification (14).

4. Mechanism of the aminoacylation reaction

The aminoacyl-tRNA synthetase reaction occurs in two steps. First, activation of the amino acid carboxyl group by ATP leads to the synthesis of an aminoacyl adenylate, in which the carboxyl group is linked to the phosphate of AMP, and elimination of pyrophosphate, which is subsequently hydrolysed to inorganic phosphate thus driving the reaction to completion. The intermediate aminoacyl adenylate usually remains firmly bound to the enzyme (I), which also catalyses the subsequent transfer of the aminoacyl group to the 3'-terminal adenosine moiety of the tRNA (II).

$$E + R\cdot CH\cdot CO_2^- + ATP \longrightarrow E\cdot R\cdot CH\cdot CO\cdot AMP + PP_i \longrightarrow R\cdot CH\cdot CO\cdot tRNA + AMP + E$$

with NH_3^+ below $R\cdot CH\cdot CO_2^-$; NH_3^+ below $E\cdot R\cdot CH\cdot CO\cdot AMP$; $2P_i$ below PP_i; NH_3^+ below $R\cdot CH\cdot CO\cdot tRNA$.

$$(I) \qquad\qquad (II)$$

Generally, the aminoacyl-tRNA synthetases use only the natural L-amino acids. An exception is D-tyrosine which can be activated by the synthetases from *E.coli* or *B.subtilis* although the V_{max} is less than 5% of that for L-tyrosine (15).

Some aminoacyl-tRNA synthetases activate the amino acid, as measured by the pyrophosphate exchange reaction (i.e. the reverse of the first reaction above), much faster than the overall aminoacylation of tRNA. In these cases, the rate-limiting step is evidently either the transfer of the aminoacyl group to the tRNA or the release of the charged tRNA from the enzyme. In other cases, however, for example, *E.coli* tyrosyl-tRNA synthetase, the rate constants are very similar. It is also noteworthy that the activation of some amino acids appears to require the presence of the cognate tRNA, suggesting that the tRNA may induce a conformational change of the synthetase (16).

In solution, the aminoacyl ester at the terminal adenosine of tRNA equilibrates rapidly between the 2'- and 3'-hydroxyl groups. Replacement of this residue by the analogues 2'- or 3'-deoxyadenosine, 2'- or 3'-aminoadenosine or adenosine in which the $C_{2'}-C_{3'}$ bond has been cleaved by periodate oxidation followed by reduction of the dialdehyde, precludes migration of the aminoacyl group and experiments with these analogues have shown that some synthetases esterify the 2'-hydroxyl group whereas others charge the 3'-position and a few charge both. Usually, prokaryotic and eukaryotic enzymes show the same specificity. In the case of tryptophan, the tRNA from *E.coli* was acylated at the 2'-hydroxyl group by both the *E.coli* and yeast synthetases, whereas the tRNA from yeast was esterified at the 3'-hydroxyl group by the same two enzymes. Thus, selection of the hydroxyl group to be acylated depends on the tRNA structure rather than the synthetases. It is also noteworthy that the analogue Phe – tRNACC-3'dA (charged phenylalanyl-tRNA with a 3' deoxyadenosine residue) is bound to ribosomes by an EF-T-dependent, GTP-hydrolysing reaction, but functions in the peptidyl-transferase reaction only as an acceptor, not as a donor. Also,

Phe-tRNA can displace the enzymically bound analogues Phe-tRNACC-3′dA and Phe-tRNACC-3′ANH$_2$ from the ribosomal A-site.

These experiments suggest that, whichever the initial acylation site of tRNA, the aminoacyl group has to be at the 3′-hydroxyl of the terminal adenosine for peptide bond synthesis. In those cases where the initial esterification occurs at the 2′-hydroxyl, a transacylation reaction is required. The chemical transacylation rates at pH 7.3 appear to be up to 10 times slower than peptide bond formation in *E.coli*, suggesting the need for an enzymic activity to catalyse the reaction in protein biosynthesis (2).

5. Accuracy of cognate aminoacyl-tRNA formation and the correction of errors

Protein synthesis takes place with a high degree of fidelity with respect to the incorporation of individual amino acids into specific positions of the polypeptide chain (17). Measurements of the utilization of valine in place of the structurally closely related amino acid isoleucine for the synthesis of rabbit globin have shown that incorrect incorporation occurs with a frequency of less than 3 in 10 000 (18).

Although valine is activated not only by its own tRNA synthetase but also by the isoleucyl-tRNA synthetase, the efficiency of the latter reaction, measured by the ratio k_{cat}/K_M, is about 200-fold lower than for isoleucine. Moreover, a proof-reading or editing mechanism enables mistakes by the synthetase to be corrected. Editing could involve either a tRNA^{Ile}-induced hydrolysis of the Val–AMP enzyme complex (I) or a post-transfer hydrolysis of the misacylated Val–tRNA^{Ile} (II) before release from the enzyme. A third possibility is that the incorrect aminoacyl adenylate complex dissociates from the enzyme faster than the transfer of the aminoacyl group to the tRNA. The relative contributions of

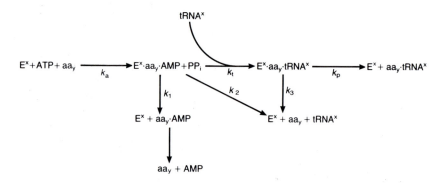

Figure 2.7. Alternative editing mechanisms to ensure accuracy of tRNA aminoacylation (19). The rate constant of the first step which leads to the formation of aminoacyl adenylate is given by k_a. The rate constant for the transfer of the amino acid residue from the aminoacyl adenylate to transfer RNA is given by k_t and that for release of free aminoacyl-tRNA from the enzyme by k_p. The editing pathways involve k_t and k_p.

these three alternative pathways has been evaluated by Fersht (19) who concluded that amino acids larger than the cognate species are sufficiently discriminated against at the activation step by large differences in k_a (*Figure 2.7*). For example, the rate of activation of isoleucine by E^{Val} is only 2×10^{-5} that of valine, which is sufficient to account for the accuracy of protein synthesis. On the other hand, amino acids with closely related (isosteric) or smaller side chains are activated at appreciable rates, although still much more slowly than the cognate species. Moreover, differences in activation rates increase with structural divergence, but editing becomes important for the prevention of errors when there is significant activation of non-cognate amino acids. Either the pre-transfer mechanism (k_1 or $k_2 \gg k_t$) or the post-transfer mechanism ($k_1 < k_t$ and $k_3 \gg k_p$) can be relevant, depending on the amino acid. Thus, the pre-transfer pathway appears to be significant for the rejection of homocysteine by E^{Met} and E^{Ile} or cysteine by E^{Ile} whereas the rejection of threonine and α-amino-isobutyrate by the valyl-tRNA synthetases of several species appears to involve hydrolysis of the mischarged tRNA (k_3).

6. Further reading

Barnett,W.E., Schwartzbach,S.D., and Hecker,L.I. (1978) The transfer RNAs of eukaryotic organelles. *Prog. Nucleic Acid Res. Mol. Biol.,* **21**, 143–79.
Björk,G.R., Ericson,J.U., Gustafsson,C.E.D., Hagervall,T.G., Jönsson,Y.H., and Wikström,P.M. (1987) Transfer RNA modification. *Annu. Rev. Biochem.,* **56**, 263–87.
Schimmel,P. (1987) Aminoacyl tRNA synthetases: general scheme of structure-function relationships in the polypeptides and recognition of transfer RNAs. *Annu. Rev. Biochem.,* **56**, 125–58.
Schimmel,P. and Söll,D. (1979) Aminoacyl-tRNA synthetases: general features and recognition of transfer RNAs. *Annu. Rev. Biochem.,* **48**, 601–48.

7. References

1. Sprinzl,M., Hartmann,T., Meissner,F., Moll,J., and Vorderwülbecke,T. (1987) *Nucleic Acids Res.,* **15**, Suppl., r53–r188.
2. Chládek,S. and Sprinzl,M. (1985) *Angew. Chem.,* **24**, 371–91.
3. Perret,V., Garcia,A., Grosjean,H., Ebel,J.-P., Florentz,C., and Giegé,R. (1990) *Nature,* **344**, 787–9.
4. Rich,A. and RajBhandary,U.L. (1976) *Annu. Rev. Biochem.,* **45**, 805–60.
5. Wilcox,M. and Nirenberg,M.W. (1968) *Proc. Natl. Acad. Sci. USA,* **61**, 229–36.
6. Natsoulis,G., Hilger,F., and Fink,G.R. (1986) *Cell,* **46**, 235–43.
7. Dang,C.V. and Dang,C.V. (1986) *Biochem. J.,* **239**, 249–55.
8. Waye,M.M.Y., Winter,G., Wilkinson,A.J., and Fersht,A.R. (1983) *EMBO J.,* **2**, 1827–9.
9. Fersht,A.R., Shi,J.-P., Wilkinson,A.J., Blow,D.M., Carter,P., Waye,M.M.Y., and Winter,G.P. (1984) *Angew. Chem.,* **23**, 467–538.
10. Rich,A. and Schimmel,P. (1977) *Nucleic Acids Res.,* **4**, 1649–65.
11. Hou,Y.-M. and Schimmel,P. (1988) *Nature,* **333**, 140–5.
12. Francklyn,C. and Schimmel,P. (1989) *Nature,* **337**, 478–81.

13. Sampson,J.R., DiRenzo,A.B., Behlen,L.S., and Uhlenbeck,O.C. (1989) *Science*, **243**, 1363 – 6.
14. Muramatsu,T., Nishikawa,K., Nemoto,F., Kuchino,Y., Nishimura,S., Miyazawa,T., and Yokoyama,S. (1988) *Nature, **336***, 179 – 81.
15. Calendar,R. and Berg,P. (1967) *J. Mol. Biol.*, **26**, 39 – 54.
16. Chapeville,F. and Rouget,P. (1972) In L.Bosch (ed.) *The mechanism of protein synthesis and its regulation.* North-Holland Publishing Co., Amsterdam, pp. 5 – 32.
17. Cornish-Bowden,A. and Wharton,C.W. (1988) *Enzyme kinetics.* IRL Press, Oxford, pp. 63 – 7.
18. Loftfield,R.B. and Vanderjagt,D. (1972) *Biochem. J.*, **128**, 1353 – 6.
19. Ferscht,A.R. (1986) In T.B.L.Kirkwood, R.F.Rosenberger, and D.J.Galas (eds) *Accuracy in molecular processes.* Chapman and Hall, London, pp. 67 – 82.

Appendix

Nomenclature of tRNAs and tRNA charging enzymes

The amino acid specificity of the tRNA in the aminoacylation reaction is indicated by a right hand superscript and the amino acid attached to the tRNA by a prefix. Thus, $tRNA^{Phe}$ represents uncharged phenylalanine-specific tRNA, Val-$tRNA^{Val}$ valine-specific tRNA charged with valine and Glu-$tRNA^{Gln}$ glutamine-specific tRNA charged with glutamic acid. The right-hand subscript position sometimes indicates the organism from which the tRNA is derived, e.g. $tRNA_{yeast}$, but has also been used for designating the anticodon, e.g. $tRNA^{Ala}_{UGC}$. Alternatively, the anticodon may be shown in the superscript after the amino acid, e.g. $tRNA^{Ala/UGC}$. Hydrogen bonding in representations of the secondary and tertiary structure is usually indicated by a single line or dot, regardless of the number of bonds in different base pairs.

The initiator tRNA, which is specific for methionine, is termed $tRNA^{Met}_{f}$ or $tRNA^{Met}_{i}$ (or sometimes $tRNA_f$ or $tRNA_i$). When charged with methionine or N-formylmethionine it is designated Met·$tRNA_f$ or fMet·$tRNA_f$ (Met·$tRNA_i$ or fMet·$tRNA_i$), respectively. The elongator methionine-specific tRNA, which inserts methionine into internal positions of the growing peptide chain, is termed Met·$tRNA_m$ when charged and $tRNA_m$ or $tRNA^{Met}_{m}$ when uncharged.

The enzymes involved in charging tRNAs were originally called amino acid activating enzymes but are now termed aminoacyl-tRNA synthetases or, less frequently, amino acid:tRNA ligases (Enzyme Commission reference number EC 6.1.1.), e.g. leucyl-tRNA synthetase or leucine:tRNA ligase.

3

Ribosome structure

1. Introduction

Ribosomes function within the physiological range of temperatures (0–80° C) and ionic conditions. For example, the range of growth temperatures is 0–20° C for psychrophilic bacteria, 20–50° C for mesophilic bacteria and above 50° C for thermophilic bacteria. The ribosomes themselves are able to function outside these limits to a certain extent, although the exact temperature range depends on the species from which they are derived. Thus, a cell-free system containing ribosomes from *Escherichia coli* (a mesophilic bacterium) and supernatant factors from *Bacillus stearothermophilus* functions over the temperature range 0–65° C. Apparently, the ribosomes from mesophilic bacteria are active over a temperature range that overlaps to a considerable extent the working range of ribosomes from psychrophilic and thermophilic bacteria. Ribosomes from mammals and amphibians have a stability roughly comparable with those from mesophilic bacteria (1).

The ionic environment of the cell in which ribosomes function is not known precisely but the potassium and magnesium ion concentrations in cells are likely to be within the range 10–200 mM and 1–10 mM, respectively. *In vitro*, mammalian ribosomes are stable in 2 M KCl, provided that the magnesium ion concentration is kept in balance (e.g. 2 M KCl/0.1 M $MgCl_2$) (2). Certain bacteria (halophiles) flourish in concentrated salt solutions (up to 4 M KCl or NaCl).

As a general rule, individual ribosomes within a population are identical except possibly for minor variations, such as the nucleotide sequence of the RNA, which do not affect function. In eukaryotes, ribosomes are found not only in the cytoplasm but also in mitochondria and chloroplasts. Prokaryotic and eukaryotic cytoplasmic and organellar ribosomes have distinctive properties (*Table 3.1*). The properties of the smaller subparticles (or small subunits) and of the larger subparticles (or large subunits) are also summarized in *Table 3.1*.

35

Table 3.1. Properties of ribosomes and ribosomal subunits

Species	$S_{20,w}$ (Svedbergs)	Size (nm) (largest dimension)	Mass (MDa)	RNA: protein ratio (w/w)	Axial ratio	rRNA Mass (MDa)	rRNA $S_{20,w}$ (nominal)
Ribosomes							
Prokaryotes							
E.coli	70	22.5 ± 2.5	2.6–2.9	2:1			
Eukaryotes							
cytoplasmic	80	28.0 ± 2.8	3.4–4.5	1:1			
chloroplast	70	22.5 ± 2.5	2.5–3.3	1:1			
mitochondrial	55–77	–	3.2–4.5	1.1:1.3			
Small ribosomal subunits							
Prokaryotes							
E.coli	30	22.0 ± 2.2	0.95	2:1	2:1	0.6	16S
Eukaryotes							
cytoplasmic	36–41	25.0 ± 2.5	1.4	1:1	2:1	0.7	18S
chloroplast	28–35	22.0 ± 2.2	1.2	1:1	2:1	0.6	16S
mitochondrial	28–40	–	1.1–1.7	1:3	–	0.3–0.7	12–14S
Large ribosomal subunits							
Prokaryotes							
E.coli	50	22.5 ± 2.2	1.75	2:1	1:1	1.17	23S
						0.04	5S
Eukaryotes							
cytoplasmic	60	28.0 ± 2.8	2.1–3.1	1:1	1:1	1.2–1.75	25–28S
						0.05	5.8S
						0.4	5S
chloroplast	46–54	22.5 ± 2.2	2.4	1:1	1:1	1.1	23S
						0.04	5S
						0.03[a]	4.5S[a]
mitochondrial	39–60	–	1.65–2.5	1:3	–	0.5–1.5	16–25S
						0.04[b]	5S[b]

[a]4.5S rRNA corresponding to the 100 nucleotides at the 3′-end of 23S rRNA in eubacteria occurs as a separate species in chloroplasts of higher plants.
[b]This species is present only in plant mitochondria and is absent from all other mitochondrial ribosomes.
– denotes data not available.

Ribosomes comprise single copies of 50 or more different proteins of M_r approximately 10 000–30 000 Da, most of which are basic, as well as multiple copies of one acidic protein some of which are chemically modified. All ribosomes contain at least two species of RNA (rRNA). A single molecule of rRNA is present within the small subunit and up to three different rRNA species are located within the large subunit. The large ribosomal subunits in the mitochondria of protozoa, fungi, and animals contain one molecule of 16–24S rRNA whereas two species (5S and 23S rRNA) are present in bacteria, plant mitochondria, and chloroplasts of algae. Chloroplasts of higher plants have 5S rRNA, 23S rRNA, and a 4.5S rRNA, which corresponds to the 100 nucleotides at the 3'-end of eubacterial 23S rRNA, and eukaryotic cytoplasmic ribosomes contain 5S rRNA, 25–28S rRNA, and a third component (5.8S rRNA) corresponding to 160 nucleotides at the 5'-end of eubacterial 23S rRNA. The 4.5S and 5.8S rRNA species are products of further steps in the maturation of rRNA precursors, which in eubacteria gives rise to 23S rRNA.

As described in Section 11 of Chapter 1, the role of the ribosome is to synthesize the polypeptide chain specified by the mRNA. Hence the ribosome has both a decoding and catalytic function. In the decoding process tRNA is accurately aligned with the mRNA so that the correct codon–anticodon interactions take place sequentially. At the end of the tRNA distal from the anticodon, the ribosomal peptidyl-transferase centre catalyses peptide bond synthesis by transfer of the nascent polypeptide chain to the next amino acid on the incoming tRNA. After this elongation of the nascent polypeptide, the mRNA is moved relative to the ribosome by one codon and the polypeptidyl tRNA is simultaneously relocated in the P-site (translocation). The large subunit of the ribosome has peptidyl transferase activity and is important for translocation, whereas the small subunit plays a major part in forming the decoding site.

Although the ribosome is usually classed as a subcellular organelle, it may also be regarded as a multi-functional multi-subunit enzyme. This enzyme is unusual because it contains RNA and originally it was thought that the ribosomal proteins were responsible for the principal activities of the ribosome while the rRNA served merely as a structural matrix. This early view has been revised and it is now more helpful for understanding the structure and function of ribosomes to regard the rRNA as important for ribosomal function with the proteins serving to enhance activities inherent in the nucleotide sequence.

A considerable effort has been devoted to structural studies in order to provide the framework essential for understanding ribosome function, including decoding of mRNA, peptidyl transferase activity, and translocation. The current status is outlined in the following sections of this chapter.

2. *Escherichia coli* ribosomes

2.1. *The importance of studies of the* E.coli *ribosome*

The belief that ribosomes, whatever their origin, fulfil the same role in protein biosynthesis and therefore have common structural features led several groups of investigators to focus on the structure of the *E.coli* ribosome using physical, chemical, biochemical, and genetic techniques. This approach has led to the elucidation of the nucleotide sequences of 16S rRNA, 5S rRNA and 23S rRNA, and of the amino acid sequences of all 52 ribosomal proteins. Furthermore, progress has been made in understanding how these components are assembled within the ribosome. Our current knowledge of the structure of the *E.coli* ribosome provides a frame of reference for studies of all other ribosomes.

2.2. The small subunit

The small subunit comprises one 16S rRNA molecule (small-subunit rRNA) and 21 distinct proteins designated S1 to S21 (see *Table 3.2*).

2.2.1. Small subunit proteins

The amino acid sequence of all 21 small subunit proteins is known (3). They range in size (see *Table 3.2*) from 8.4 kDa to 61 kDa (average 16 kDa). Most proteins are globular with an average content of 28% α-helix and 20% β-structure. Proteins S1, S2, and S6 are acidic whereas the remainder are basic. The interactions between the negatively-charged rRNA and the basic proteins contribute to the stability of the ribosome.

2.2.2. 16S rRNA

The primary and secondary structures of *E.coli* 16S rRNA (4) are presented in *Figure 3.1*. Ten of the 1542 nucleotides are modified by methylation. Ribosomes constructed from non-methylated 16S rRNA are able to function in *in vitro* protein synthesis but with diminished efficiency. On the other hand, the absence of m_2^6A-1518 and m_2^6A-1519 modifications renders the ribosome resistant to the antibiotic kasugamycin.

The 16S rRNA component of the small subunit plays an active role in the initiation, elongation, and termination steps of protein biosynthesis (5). For example, the anti-Shine – Dalgarno sequence participates in binding mRNA by base pairing with the Shine – Dalgarno sequence located near to, but upstream from, the AUG start codon of the mRNA (see Section 2.1 of Chapter 4). This interaction is probably also important for maintaining mRNA in the correct reading frame (8).

Table 3.2. The number and size of *E.coli* small subunit proteins (3)

Protein	Number of amino acid residues	Mass (Da)	Protein	Number of amino acid residues	Mass (Da)
S1	557	61 159	S2	240	26 613
S3	232	25 852	S4	203	23 137
S5	166	17 515	S6	135	15 704
S7B	153	17 131	S7K	177	19 732
S8	129	13 996	S9	129	14 725
S10	103	11 736	S11	128	13 728
S12	123	13 606	S13	117	12 968
S14	98	11 191	S15	88	10 137
S16	82	9 191	S17	83	9 573
S18	74	8 896	S19	91	10 299
S20	86	9 553	S21	70	8 369

The nomenclature is based on the positions of the proteins in two-dimensional polyacrylamide gel electrophoresis. Protein S7 is different in K-strains compared with the B-strain.

The decoding site, where the codon–anticodon interaction takes place, lies within 1 or 2 nm of a C residue at position 1400. The 5′ (wobble) base of certain tRNA anticodons can be cross-linked by UV light with this C residue when the tRNA is in the P-site. The highly conserved nucleotide sequences around positions 1400 and 1500 help to form the decoding site. These and other regions implicated in termination and subunit association are summarized in *Figure 3.1*. The functional regions of 16S rRNA are probably located on the surface of the ribosome. This expectation has been confirmed in seven cases by DNA hybridization electron microscopy (7) using biotinylated oligodeoxyribonucleotides to locate complementary regions in the small subunit, the hybridized probe being detected by its affinity for avidin.

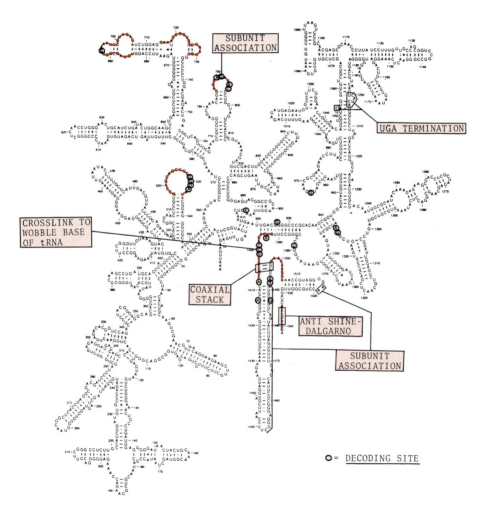

Figure 3.1. Functional regions of *E.coli* 16S rRNA (5). The scheme for the secondary structure is that of Noller *et al.* (6). The coloured regions are exposed on the surface of the small subunit, as shown by DNA hybridization electron microscopy (7).

2.2.3. Phylogenetic importance of 16S rRNA

Methods developed for the rapid sequencing of DNA have provided a large
database comprising the primary structure of rRNA genes of both bacteria and
eukaryotes, including chloroplasts and mitochondria. The data amply confirm
the early notion that certain features of rRNA are more highly conserved than
others (9). The identification of highly conserved sequences has led to an
extension of the database by direct sequencing of rRNA using reverse
transcriptase and appropriate oligodeoxyribonucleotide primers as well as by
amplifying and sequencing rRNA genes. These studies have shown that 16S
rRNA sequences follow a common pattern of secondary structure (see *Figure
3.1*) and have led also to new insights into the phylogeny of bacteria. Comparison
of 16S rRNA sequences is now a widely accepted method for establishing
phylogenetic relationships (10).

2.2.4. The *in vitro* assembly of the small subunit from 16S rRNA and ribosomal proteins

The small ribosomal subunit is capable of self-assembly *in vitro* from the 16S
rRNA and ribosomal proteins. Nomura and his colleagues studied the function
of proteins by first taking the small subunit apart (dis-asscmbly), then re-
assembling (reconstructing or reconstituting) the subunit under controlled
conditions and scrutinizing its function (11). Hybrid ribosomes assembled *in vitro*,
for example from *E.coli* 16S rRNA and *B.stearothermophilus* proteins (or *vice versa*)
were shown to be fully functional in protein biosynthesis (12) illustrating the
highly conserved nature of the functional regions of the small subunit.

The *in vitro* assembly of the small subunit follows the pathway illustrated in
Figure 3.2 which shows that six S proteins (S4, S7, S8, S15, S17, and S20) bind
directly and independently to rRNA. Subsequently, the remaining proteins
complete the assembly process.

In vitro self-assembly has been used to establish the spatial arrangement of
the proteins within the small subunit (13). To do this, bacteria were grown in
D_2O so that the proteins of the small subunit were labelled with deuterium.
After isolating the individual proteins, the subunits were reconstructed from the
rRNA and proteins, replacing two of the proteins with proteins which were
labelled with deuterium. Finally, the separation within the small subunit of the
centres of mass of the pair of deuterated proteins was measured by neutron
scattering. From a series of such experiments using different pairs of deuterated
proteins a complete three-dimensional map was constructed showing the relative
locations of the proteins (*Figure 3.3*).

2.2.5. Models of the small ribosomal subunit

Models serve a useful role in summarizing a wealth of data thereby bringing
particular features into sharper focus, and they are transitional steps towards
establishing a definitive picture. A consensus view of the outline of the small
subunit in conventional electron microscopy (see Section 2.4.1) is shown in *Figure*

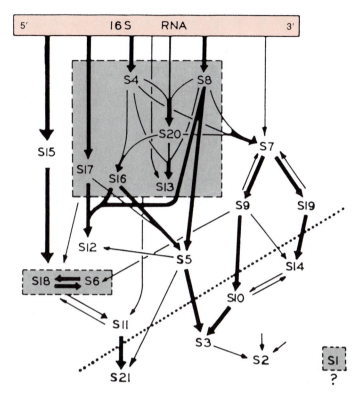

Figure 3.2. Assembly map of the *E.coli* small ribosomal subunit (11). Arrows between proteins indicate a facilitating effect of one protein on another in the *in vitro* assembly of 30S subunits; a thick arrow indicates a major effect. Proteins above the dotted line are required for the formation of active reconstitution (RI*) particles.

3.4 a and *b*; the binding sites for initiation factors IF-1, IF-2, and IF-3, and elongation factors EF-Tu are indicated in *Figure 3.4 c* and *d*. *Figure 3.4 e* is a summary map of the approximate locations of the 5′- and 3′-ends of 16S rRNA and of several proteins. The platform is the likely site of the codon–anticodon interaction. This region includes the 3′-end of the 16S rRNA, the N^6-dimethyladenine nucleotides 1518 and 1519 and the initiation codon of mRNA preceded by a Shine–Dalgarno sequence. Proteins S6, S11, S15, and S18 are also found on the platform and are implicated in the binding of mRNA. Proteins S3, S10, S14, and S19 are involved in the poly(U)-dependent binding of tRNA[Phe].

The way in which 16S rRNA and proteins interact to form the small subunit is now known in outline, based on the relative distribution of proteins (see *Figure 3.3*) and a knowledge of the regions of 16S rRNA to which the individual proteins bind. The binding sites of particular proteins were investigated by UV-crosslinking experiments (15) which result in covalent bond formation between a nucleotide and an amino acid in close proximity. Twelve of the 21 possible binding

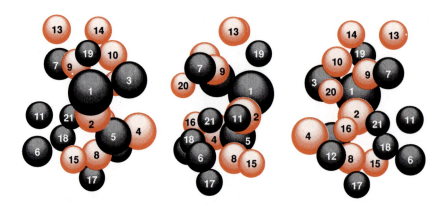

Figure 3.3. The spatial arrangement of proteins in the small ribosomal subunit from *E.coli* (13). The model of the placement of proteins, derived from neutron scattering measurements, is shown from the cytoplasmic side (left), edge on (middle) and the subunit interface side (right). For convenience, proteins are depicted as spheres drawn to scale whose centres are at positions determined for protein centroids by scattering data. The proteins are identified by numbers (see *Table 3.2*). The array is approximately 19 nm from the top of S13 to the bottom of S17.

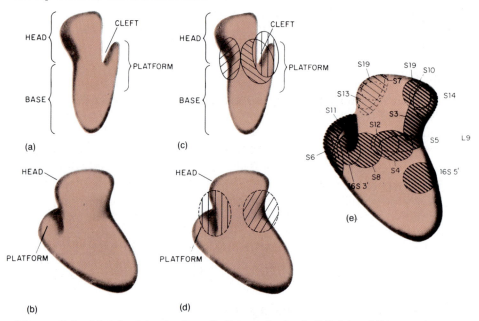

Figure 3.4. Models of the *E.coli* small ribosomal subunit (14). **(a)** and **(b)** are consensus models of the view of the small subunit in conventional electron microscopy showing the principal features of the head, cleft, and platform; **(c)** and **(d)** shown the binding sites for initiation factors IF-1, IF-2 and IF-3 and elongation factors EF-G and EF-Tu; **(e)** is a summary map of the location of the 3'- and 5'-ends of 16S rRNA and several small subunit proteins.

sites were identified in this way. A complete map of all 21 sites was obtained by using kethoxal as a probe for changes in the chemical reactivity of guanine residues which correlate with protein binding (16). The successive binding of proteins to 16S rRNA in the *in vitro* assembly of the small subunit was investigated by the kethoxal technique. The results of this chemical protection method agree with the UV-cross-linking data in the dozen cases where a comparison is possible. The model obtained by fitting the 16S rRNA sequences to proteins is presented in *Figure 3.5*.

2.3. The large subunit

The peptidyl transferase and translocase activities of the ribosome are properties of the large subunit which comprises 5S and 23S rRNAs as well as 31 different proteins. The greater complexity of the large subunit compared with the small subunit has caused knowledge of its structure to lag behind that of the small subunit.

2.3.1. Large subunit proteins

Twenty-nine of the 31 proteins of the large subunit are present as a single copy per subunit and most are basic. However, at least two copies of the acidic proteins L7 and L12 are present. These proteins have the same amino acid sequence but differ in the amino terminus which is *N*-acetylserine in L7 and serine in L12. The molecular sizes of the proteins are given in *Table 3.3*. The amino acid sequences of all 31 proteins have been established (3).

2.3.2. 23S rRNA and 5S rRNA

The 23S rRNA component comprises 2904 nucleotides (17) which fold into a partly bihelical structure (18), features of which have been conserved throughout evolution (9). As with 16S rRNA the secondary structure of *E.coli* 23S rRNA serves as a model for all other homologous rRNA species. The secondary structure and important functional regions (19) of 23S rRNA are illustrated in *Figure 3.6*. Sequences of 23S rRNA implicated in forming parts of the A-, P- and E-sites are shown in detail in *Figure 3.7*.

The peptidyl transferase site was sought on the basis of the assumption that the −CCA 3'-end of tRNA would base pair with its complementary 5'UGG3' sequence within 23S rRNA. A single sequence at positions 803−811 was found to be highly conserved among all 23S rRNA species and this 5'UAGCUGGUU3' sequence proved to be the only possibility (20). Tertiary folding of 23S rRNA suggests that the peptidyl transferase site comprises the two regions centred around positions 803−811 and 2030−2615 (19,20,27).

Elongation factors EF-Tu and EF-G compete with each other for a binding site on the ribosome and EF-G cross-links to 23S rRNA in the region of position 1067. Moreover, methylation of the ribose of this nucleotide residue confers resistance to inhibition of the interaction of EF-G with ribosomes by the antibiotic thiostrepton. Chemical protection studies suggest that EF-G also interacts with

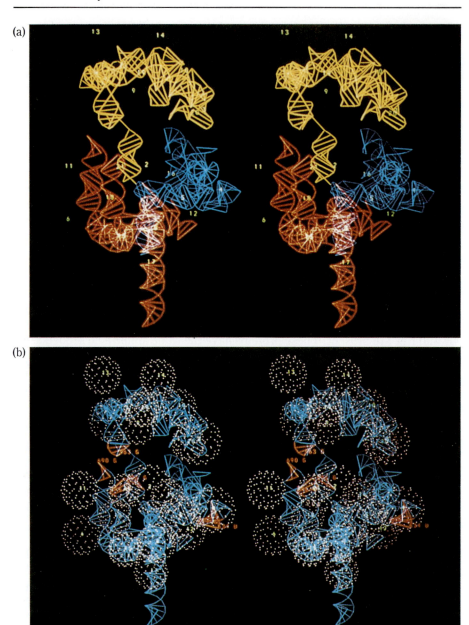

Figure 3.5. Model showing structural domains of the 16S rRNA within the *E.coli* small ribosomal subunit (16). **(a)** Stereo pair viewed from the side away from the 30S – 50S subunit interface, showing the 5′, central and 3′ major domains in blue, red and yellow, respectively. The numbers indicate the positions of the centres of individual ribosomal proteins; **(b)** as in a except that the tRNA-protected regions around positions 693, 790, 926, and 966 (upper left), the 530 loop (far right), and the streptomycin and spectinomycin-protected regions around positions 910 and 1064, respectively, are shown in red.

Table 3.3. The number and size of *E.coli* large ribosomal subunit proteins (3)

Protein	Number of amino acid residues	Mass (Da)	Protein	Number of amino acid residues	Mass (Da)
L1	233	24 599	L2	272	29 730
L3	209	22 258	L4	201	22 087
L5	178	20 171	L6	176	18 832
L7 (2 copies)	120	12 220	L8 (= L7+L10)	–	–
L9	148	15 696	L10	164	17 581
L11	141	14 874	L12 (2 copies)	120	12 178
L13	142	16 019	L14	123	13 541
L15	144	14 981	L16	136	15 296
L17	127	14 365	L18	117	12 770
L19	114	13 002	L20	117	13 366
L21	103	11 565	L22	110	12 227
L23	100	11 209	L24	103	11 185
L25	94	10 694	L26 (= S20)	–	–
L27	84	8 993	L28	77	8 875
L29	63	7 274	L30	58	6 411
L31	62	6 971	L32	56	6 315
L33	54	6 255	L34	46	5 381

Each protein is present as one copy per ribosome except for L7 and L12. The nomenclature of the proteins is based on their separation by two-dimensional gel electrophoresis. L7 is identical to L12 except that its amino terminal serine residue is acetylated. Excluding L7, L8, and L26 there are 31 distinct proteins.

the highly conserved nucleotide sequence near position 2660. This position is the site of action of the cytotoxins α-sarcin and ricin. Except for position 1067 EF-Tu protects the same region as EF-G (19,21).

5S rRNA comprises 120 nucleotides. The sequence near the 5′-end is complementary to that near the 3′-end, giving rise to a Y-shaped secondary structure (22). There appears to be no further folding, suggesting the absence of tertiary interactions (23).

Although 5S rRNA is essential for ribosome activity its role is unknown. It is likely, however, that it is involved in an interaction with 16S rRNA during ribosome function, by analogy with known interactions of 5S rRNA and 18S rRNA in eukaryotes (24). Whether this putative interaction takes place during the decoding or translocation step or both is still unclear. In the assembly process, proteins L18 and L25 bind to 5S rRNA co-operatively with protein L5 prior to interaction with 23S rRNA (see *Figure 3.8*).

2.3.3. RNA – protein interactions

The large subunit, like the small one, may be assembled *in vitro* from 23S rRNA, 5S rRNA, and the large subunit proteins (25). An 'assembly map' comparable with that shown in *Figure 3.2* has been established (26) and the re-assembly process has been used to incorporate pairs of deuterated proteins in experiments

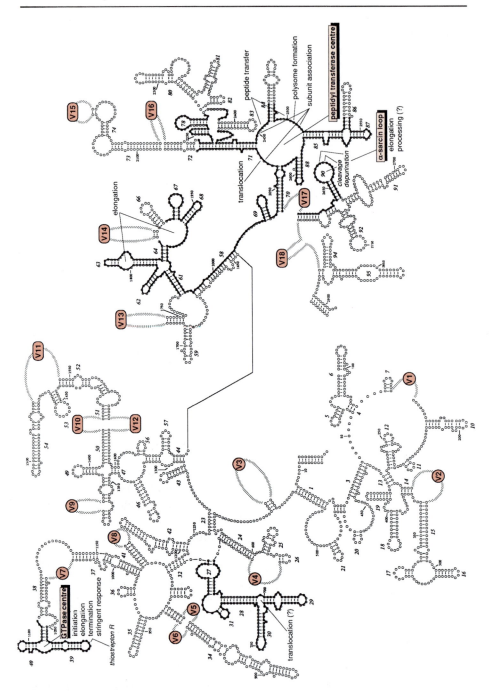

Figure 3.6. Functional regions of 23S rRNA (19). The secondary structure is shown schematically and represents large subunit rRNA in general. Expansion (or variable) segments are numbered sequentially from the 5′-end. Functional regions are labelled. Filled circles denote generally conserved regions.

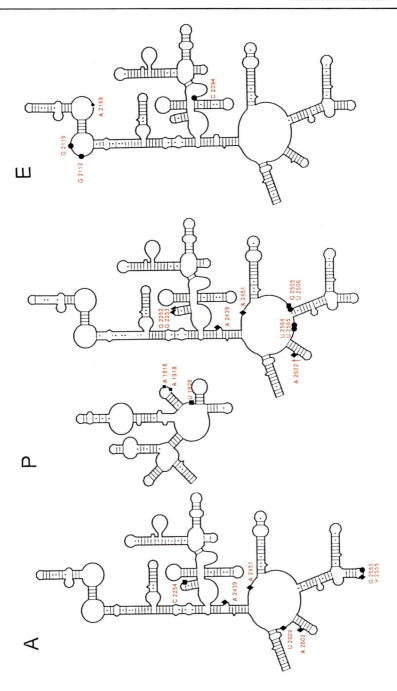

Figure 3.7. Sites of protection of 23S rRNA by tRNA in the A-, P-, and E-sites (21). Symbols show dependence of protection on the acyl moiety (◆), the 3′-terminal A (●), the 3′-terminal CA (▲) and the rest of the tRNA molecule (■).

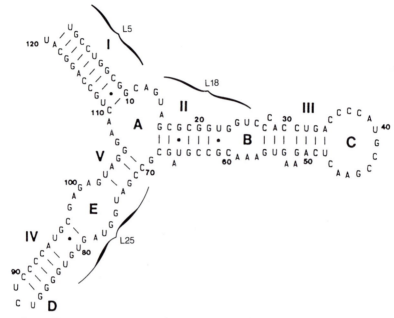

Figure 3.8.   Secondary structure of mature *E.coli* 5S rRNA from the *rrnB* operon (22). Base pairing is indicated by lines (Watson – Crick base pairs) or dots for G-U base pairs proved phylogenetically. The diagram shows helical segments (I – V), single-strand loops (A – E), and contacts with proteins (L5, L18, and L25).

designed to determine the relative positions of the proteins. The locations of several proteins with respect to 23S rRNA sequences have been identified by UV-cross-linking experiments. The data on protein – RNA interactions, the tertiary structure of 23S rRNA, and protein – protein interactions have been summarized in a model of the large subunit (27).

2.3.4. Models of the large ribosomal subunit

Conventional electron microscopy reveals the large subunit to have a distinctive contour. The roundish body has three features known as the (L1) ridge, the central protuberance and the (L7/L12) stalk. The central protuberance comprises protein L27, protein L18, and 5S rRNA. The stalk consists of two L7/L12 dimers with L10 at the base (*Figure 3.9 a* and *b*). The peptidyl transferase activity is located between the ridge and the central protuberance. The stalk is implicated in translocation. The exit site for the nascent polypeptide is opposite the P-site and the membrane-binding site is located in the region of protein L19 (*Figure 3.9c*). The arrangement of the proteins within the subunit is summarized in *Figure 3.9 d* and *e*.

2.4. Ultrastructural studies

The *E.coli* ribosome is highly asymmetric because it comprises three different rRNA components (4566 nucleotides) and 53 different proteins (approximately

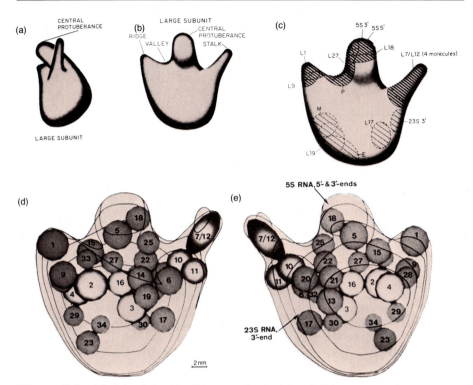

Figure 3.9. Models of the *E.coli* large subunit. (a) and (b) represent the consensus view derived from conventional electron microscopy (see *Section 2.4.1*), showing the central protuberance, the L1 ridge, and the L7/L12 stalk (14). (c) is a summary map giving the location of several large subunit proteins, 5S rRNA, 23S rRNA, and functional regions. P, M, and E refer to the peptidyl transferase site, the membrane-binding site, and the exit site for nascent protein, respectively (14). The L7/L12 stalk is implicated in translocation. (d) and (e) show the arrangement of large subunit proteins within the large subunit (27).

9000 amino acids). The lack of symmetry has hindered structural studies because there are no regular or repetitive features which could be used for image enhancement by the established techniques of superposition and rotation. Also, ribosomes are difficult to visualize by electron microscopy because of their comparatively small size (see *Table 3.1*), their high degree of hydration, sensitivity to electron beam damage, and their low contrast. However, as discussed in the next sections, a general measure of agreement has been reached on the appearance of the *E.coli* ribosome and its subunits (for models see *Figures 3.4* and *3.9*).

2.4.1. Conventional electron microscopy of *E.coli* ribosomes and subunits

E.coli ribosomes appear as round particles with a diameter of 22.5 ± 2.5 nm. The small subunit is prolate with dimensions of 22.0 × 10.0 nm ± 10%. The

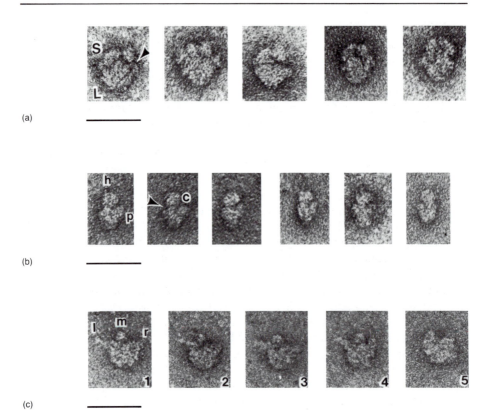

Figure 3.10. Electron micrographs of *E.coli* ribosomes and subunits (29). Conventional electron micrographs, in which the bar represents 25 nm, show: **(a)** a gallery of ribosomes revealing the mutual orientation of the small subunit (S) and the large subunit (L) with the arrow pointing to the interface. **(b)** A gallery of small subunits orientated in various views, h denoting the head, p the side protrusion (platform), c the cleft between the head and side protrusion, the arrow pointing to the one-third partition below the head. **(c)** A gallery of large subunits with the characteristic features of the crown, the central protuberance (m), the L7/L12 stalk (1) and the ridge (r). The last frame (5) in the gallery shows the crescent view; the crown view is seen much more commonly.

principal features include a head, a one-third partition, and a side protrusion. The larger subunit has dimensions of 22.5 nm $\pm$ 10% with an apical crown comprising three crests, the middle crest being more rounded than the two uneven side crests. Representative views of the ribosome and its subunits are presented in *Figure 3.10*.

2.4.2. Immuno-electron microscopy

The models presented in *Figure 3.9* were obtained by a critical (but nevertheless subjective) visual examination of images, but additional information is needed to obtain greater resolution. The asymmetry of the ribosome has been exploited by the application of immuno-electron microscopy which detects antibodies raised to a particular ribosomal protein after reaction with ribosomes. The Y-shaped antibody may react with one or two ribosomes (or ribosomal subunits) and the orientation of the small subunit with respect to the large subunit has been established in double-labelling experiments, using one antibody specific for a particular small subunit protein and another with specificity for a protein of the large subunit. The model derived by this approach is illustrated in *Figure 3.11*.

2.4.3. Quantitative reconstruction of the three-dimensional structure from single particle images (30,31)

The image of the ribosome as seen in the electron microscope is a two-dimensional projection of the actual three-dimensional structure onto a plane perpendicular to the electron beam. A technique termed the random-conical-tilt series method has been developed to produce the three-dimensional structure from the two-dimensional image of the asymmetrical ribosome. The analysis assumes that the principal structural features of biologically active ribosomes or subunits are preserved after the sample is bound to the specimen support film and after irradiation with electrons. A further minimum requirement is that the major structural features should be independent of the orientation of the specimen with respect to the support. This appears to be the case. Although ribosomes are asymmetric, an objective analysis of electron microscopic images may be made if the ribosomes sit on the grid in a limited number of preferred orientations. In such a case, the optical density of the image may be measured by microdensitometry and the data stored in a computer. If, say, 50 images, each in the same orientation are processed, the combined data give a fifty-fold amplified signal whereas the noise is averaged out. The composite image thus obtained is then further analysed by image processing.

In the random-conical-tilt method only one image of a particle is obtained at a high-tilt angle, although the image at 0° tilt is used for reference. A subset of particles that show the same appearance on the specimen support film is then identified by multivariate statistical analysis and the three-dimensional structure is deduced from the composite image. This methodology also leads to a much higher resolution (see *Figure 3.12*) than is obtainable by conventional electron microscopy (*Figure 3.10*), revealing new features such as the interface canyon on the surface of the large subunit which interacts with the small subunit. The interface canyon is a deep trough located just below the central protuberance, and which extends across the subunit from the L1 ridge to the base of the L7/L12 stalk. This canyon appears to contain at least three subregions or pockets; the middle one of these is located near the peptidyl transferase centre and is actually a hole which leads to the back of the subunit. A channel may connect this pocket with the putative exit site of the nascent polypeptide chain (see *Figure 3.11b*), identified by immuno-electron microscopy. Features of the interface canyon seen from the view presented in *Figure 3.12b* are illustrated in the model shown in *Figure 3.13a*. The space between the interfaces of the two subunits (see *Figures 3.12b* and *3.13b*) is sufficient to accommodate both aminoacyl-tRNA and peptidyl-

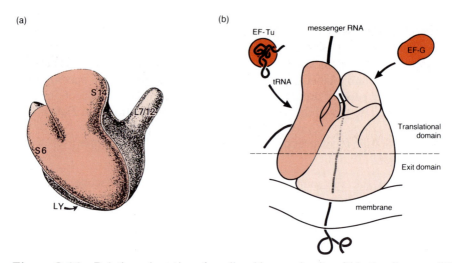

Figure 3.11. Relative orientation of small and large subunits within the ribosome (28). (a) Location of proteins S6, S14, L7/L12, and LY identified in the double-labelling experiment. (b) Model showing the relative orientation of the subunits and other functional regions (14).

tRNA (*Figure 3.13c*).

2.4.4. Three-dimensional structure of the ribosome based on electron microscopy of two-dimensional arrays (32)

In response to changes in environmental conditions such as exposure to cold, a cell may have to store potentially active ribosomes for long periods. In these situations, some ribosomes form ordered arrays (two-dimensional sheets) which are useful for study by electron microscopy because the results may be analysed by three-dimensional image reconstruction techniques. It is now possible to grow two-dimensional sheets of ribosomes and ribosomal subunits *in vitro*. The best results, yielding models at resolutions of 4.7 and 3.0 nm, respectively, have been obtained with *B.stearothermophilus* ribosomes and their large subunits. Image reconstruction from electron micrographs of ribosomes shows structures very similar to those described in Section 2.4.1, including a cleft in the large subunit which merges into a hole or tunnel, approximately 2.5 nm in diameter, extending for 10 – 12 nm (cf. *Figure 3.12*).

2.4.5. X-ray crystallographic studies (33)

Whereas ribosomes are rather small for electron microscopy, they are huge as objects for X-ray crystallography, but analysis is complicated by their size, complexity, and asymmetry, as previously discussed.

   The first step in crystallographic studies is the preparation of suitable crystals: methods have been devised to crystallize both ribosomes and ribosomal subunits

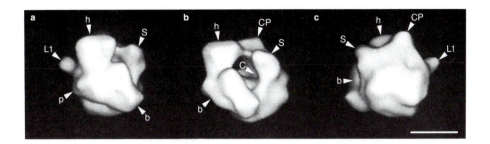

Figure 3.12. Reconstructed images of the *E.coli* ribosome presented from three viewing directions (34,35). The bar represents 10 nm. (**a**) Ribosome presenting Lake's asymmetric view (14) with the small subunit in front. (**b**) View after 40° rotation from (**a**) around an axis lying vertically in the plane of the image. (**c**) View after 150° rotation from (**a**) around the same axis. Small subunit: p, platform; h, head; b, main body. Large subunit: L1, L1 ridge; CP, central protuberance; S, stalk base (the stalk is not visible in the merged reconstruction because of its high mobility); C, interface canyon, which runs the entire width of the subunit perpendicular to the plane of the figure.

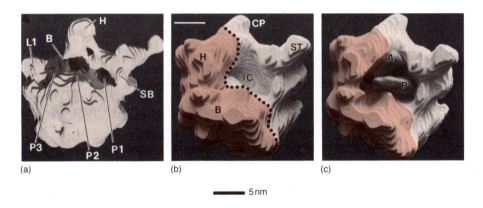

Figure 3.13. Styrofoam models representing the three-dimensional structure of *E.coli* large subunits and ribosomes (compare *Figure 3.12b*). (**a**) Large subunit (30): L1, the L1 ridge; B, the flexible bridge between L1 and H (the central protuberance); SB, stalk base (the L7/L12 stalk extends from SB); P1, P2, and P3, pockets of the interface canyon. (**b**) Model of the ribosome (31). The dashed line indicates the boundary between the small and large subunits. Small subunit: H, head; B, body. Large subunit: CP, central protuberance; ST, L7/L12 stalk; IC, interface canyon. (**c**) Ribosome with scale models of two tRNA molecules inserted into the interfacial space. A, aminoacyl tRNA; P, peptidyl tRNA. The bar denotes 5 nm.

in vitro. A major practical difficulty has been the instability of preparations due to trace contaminants such as proteases which cause slow breakdown and inactivation. Ribosomes from organisms accustomed to extreme conditions, namely the archaebacteria *B.stearothermophilus* and *Halobacterium maris mortui* (a halophile stable in concentrated salt solutions) have so far proved to be the most suitable for preparing crystals which diffract to 1.5–1.8 nm resolution. Work is in progress to derive phase information and when these data become available our view of ribosome structure will gain in precision.

2.4.6. Conclusion

The structural studies of *E.coli* ribosomes and ribosomal subunits have provided a frame of reference for understanding the biochemical mechanisms which underlie the translation of mRNA by stepwise decoding and peptide bond formation. The results of these investigations have been used to construct a model which is thought to be applicable also to all other ribosomes.

3. Further reading

Hardesty,B. and Kramer,G. (1986) (eds) *Structure, function and genetics of ribosomes*. Springer-Verlag, Berlin.

Hill,W.E., Dahlberg,A., Garrett,R.A., Moore,P.B., Schlessinger,D., and Warner,J.R. (1990) (eds) *The ribosome, structure, function and evolution*. American Society for Microbiology, Washington DC.

Liljas,A. (1991) Comparative biochemistry and biophysics of ribosomal proteins. *Int. Rev. Cytol.*, **124**, 103–36.

Nomura,M., Tissières,A., and Lengyel,P. (1974) (eds) *Ribosomes*. Cold Spring Harbor Laboratory Press, Cold Spring Harbor, NY.

Sarma,R.H. and Sarma,M.H. (1988) (eds) *From proteins to ribosomes*, Volumes 1 and 2. Adenine Press, Guilderland, New York.

Mechanism and control of translation. (1990) A collection of papers presented by many different authors at a meeting in Noordwijkerhout, The Netherlands, on 12–17 May 1990. *Biochim. Biophys. Acta*, **1050**, 1–360.

Sequences supplement. (1990) A compilation of nucleic acid sequences. *Nucleic Acids Res.*, **18S**, 2215–588.

4. References

1. Cox,R.A., Pratt,H., Huvos,P., Higginson,B., and Hirst,W. (1973) *Biochem. J.*, **134**, 775–93.
2. Cox,R.A. and Hirst,W. (1976) *Biochem. J.*, **160**, 521–31.
3. Wittmann-Liebold,B. (1984) *Adv. Protein Chem.*, **36**, 56–78.
4. Brosius,J., Dull,T.J., Sleeter,D.D., and Noller,H.F. (1981) *J. Mol. Biol.*, **148**, 107–27.
5. Dahlberg,A.E. (1989) *Cell*, **57**, 525–9.
6. Noller,H.F., Stern,S., Moazed,D., Powers,T., Svenson,P., and Changchien,L.-M. (1988) *Cold Spring Harbor Symp. Quant. Biol.*, **52**, 695–708.
7. Oakes,M.I. and Lake,J.A. (1990) *J. Mol. Biol.*, **211**, 897–906.
8. Weiss,R.B., Dunn,D.M., Atkins,J.F., and Gesteland,R.F. (1987) *Cold Spring Harbor Symp. Quant. Biol.*, **52**, 687–93.
9. Raué,H.A., Klootwijk,J., and Musters,W. (1988) *Progr. Biophys. Mol. Biol.*, **57**, 77–129.
10. Woese,C.R. (1987) *Microbiol. Rev.*, **51**, 221–71.

11. Nomura,M. (1973) *Science,* **179**, 864 – 93.
12. Nomura,M., Traub,P., and Bechmann,H. (1968) *Nature,* **219**, 793 – 9.
13. Moore,P.B. (1988) *Nature,* **331**, 223 – 7.
14. Oakes,M.I., Scheinman,A., Atha,T., Shankweiler,G., and Lake,J.A. (1990) In W.E.Hill *et al.* (eds) *The ribosome, structure, function and evolution.* American Society for Microbiology, Washington, DC, pp. 180 – 93.
15. Brimacombe,R., Atmadja,J., Stiege,W., and Schuler,D. (1988) *J. Mol. Biol.,* **199**, 115 – 36.
16. Stern,S., Powers,T., Changchien,L.-M., and Noller,H.F. (1989) *Science,* **244**, 783 – 90.
17. Noller,H.F. (1984) *Annu. Rev. Biochem.,* **53**, 119 – 62.
18. Noller,H.F., Kop,J., Wheaton,V., Brosius,J., Guttell,R.R., Kopylov,A., Dohme,F., Herr,W., Stahl,D.A., Gupta,R., and Woese,C.R. (1981) *Nucleic Acids Res.,* **9**, 6167 – 89.
19. Raué,H.A., Musters,W., Rutgers,C.A., Riet,J.V., and Planta,R.J. (1990) In W.E.Hill *et al.* (eds) *The ribosome, structure, function and evolution.* American Society for Microbiology, Washington, DC, pp. 217 – 35.
20. Barta,A., Steiner,G., Brosius,J., Noller,H.F., and Kuechler,E. (1984) *Proc. Natl. Acad. Sci. USA,* **81**, 3607 – 11.
21. Moazed,D. and Noller,H.F. (1989) *Cell,* **57**, 585 – 97.
22. Egebjerg,J., Christiansen,R.S., Larsen,N., and Garrett,R.A. (1989) *J. Mol. Biol.,* **206**, 651 – 68.
23. Westhof,E., Romby,P., Romaniuk,P.J., Ebel,J.-P., Ehresmann,C., and Ehresmann,B.(1989) *J. Mol. Biol.,* **207**, 417 – 39.
24. Azad,A.A. (1979) *Nucleic Acids Res.,* **7**, 1913 – 29.
25. Nomura,M. and Erdmann,V.A. (1979) *Nature,* **228**, 744 – 8.
26. Rohl,R., Roth,H.E., and Nierhaus,K.H. (1982) *Hoppe-Seyler's Z. Physiol. Chem.,* **363**, 143 – 57.
27. Brimacombe,R., Gornicki,P., Rinke-Appel,J., Schuler,D., and Stade,K. (1990) *Biochim. Biophys. Acta,* **1050**, 8 – 13.
28. Lake,J.A. (1982) *J. Mol. Biol.,* **161**, 89 – 106.
29. Boublik,M. (1987) *Int. Rev. Cytol.,* **17** Suppl., 357 – 89.
30. Carazo,J.M., Wagenknecht,T., Radermacher,M., Mandiyan,V., Boublik,M., and Frank,J. (1988) *J. Mol. Biol.,* **201**, 393 – 404.
31. Wagenknecht,T., Carazo,J.M., Radermacher,M., and Frank,J. (1989) *Biophys. J.,* **55**, 455 – 64.
32. Yonath,A. and Wittmann,H.G. (1988) In R.H.Sarma and M.H.Sarma (eds). *From proteins to ribosomes.* Adenine Press, Guilderland, New York, pp. 191 – 207.
33. Hansen,H.A.S., Volkmann,N., Piefke,J., Glotz,C., Weinstein,S., Makowski,I., Meyer,S., Wittmann,H.G. and Yonath,A. (1990) *Biochim. Biophys. Acta,* **1050**, 1 – 7.
34. Penczek,P. and Frank,J. (1991) *Ultramicroscopy,* in press.
35. Frank,J., Penczek,P., Grassucci,R., and Srivastava,S. (1991) *J. Cell Biol.,* **115**, 597 – 605.

4

The ribosome cycle and translation of the genetic message

1. Introduction

Translation of the genetic information encoded in the base sequence of mRNA into the amino acid sequence of polypeptides takes place on ribosomes by stepwise formation of peptide bonds, starting from the amino terminal residue and terminating when the last amino acid has been added at the carboxyl end. This process involves the binding of a ribosome to an initiation sequence of nucleotides near the 5′-end of mRNA, movement of the ribosome along the messenger by translocation through a distance of three bases (one codon) after each peptide bond has been synthesized, and finally release of both the completed polypeptide chain and the ribosome when the termination codon at the 3′-end of the mRNA coding sequence is reached. The ribosomal subunits that are generated by dissociation of ribosomes at the end of each ribosome cycle are available for another round of protein synthesis by binding to initiation sites on free mRNA or to mRNA to which ribosomes are already attached (*Figure 4.1*). Kinetic measurements *in vitro* indicate that initiation by the first ribosome on free messenger may differ in some way from reinitiation (1).

Most mRNA in the cytoplasm is associated with one or several ribosomes in complexes known as polyribosomes or polysomes, and the pool of ribosomal subunits is small. The number of ribosomes in a polysome is governed by the length of the mRNA coding sequence and by the relative rates of initiation, elongation, and termination in the ribosome cycle. In the case of polysomes synthesizing globin chains, the coding sequence of the mRNA contains about 450 nucleotides and the most abundant size of polysome is the pentamer, indicating an average centre-to-centre distance between consecutive ribosomes of some 90 nucleotides of mRNA under normal conditions. If initiation of protein synthesis is slow relative to elongation and termination, ribosomes will be more widely spaced along the mRNA, resulting in small polysomes; conversely, rapid initiation compared with elongation and/or termination will result in closely

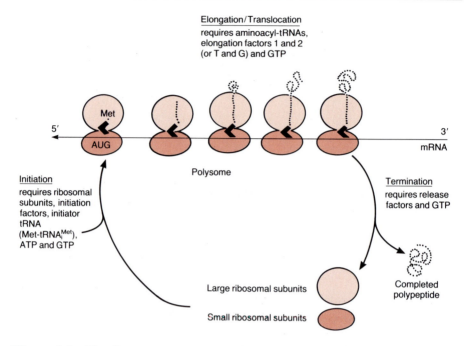

Figure 4.1.   The ribosome cycle. The arrow indicates the direction of messenger RNA movement relative to ribosomes during each translocation step. ◄ symbolizes transfer RNA and the nascent polypeptide chain is indicated by the dotted line.

packed ribosomes on the mRNA giving rise to large polysomes. Usually, the efficiency of translation of a particular mRNA is determined by the initiation rate, and control of initiation is an important factor in the regulation of protein synthesis. When initiation is inhibited, for example by the action of certain antibiotics such as aurintricarboxylic acid or pactamycin, ribosomes already engaged in protein synthesis are able to complete polypeptide chain elongation, termination and dissociation from the messenger (run-off).

   This chapter is concerned with the molecular mechanisms involved in the different stages of the ribosome cycle and the participation of a large number of different protein factors (see *Tables 4.1 – 4.3*) together with GTP and ATP in the production of polypeptide chains.

2.   Formation of the initiation complex

2.1. Structure and function of mRNA

All mRNAs are single-stranded polyribonucleotides containing a central coding region, which determines the amino acid sequence of the gene product, together with flanking sequences adjacent to the initiation and termination codons (*Figure 4.2*).

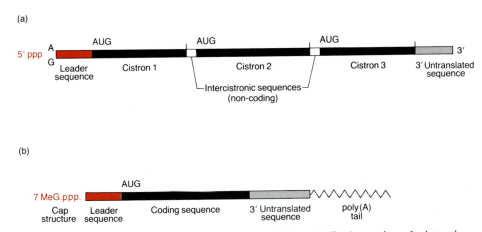

Figure 4.2. Structures of typical messenger RNAs. **(a)** Prokaryotic polycistronic mRNA. **(b)** Eukaryotic mRNA.

Within the coding region of mRNA, secondary structure appears to have little influence on the translation rate. In polycistronic mRNAs, secondary structure can regulate the frequency with which individual cistrons are read. The secondary structure of the 5'- and 3'-untranslated sequences may contain features important for the translational control of the synthesis of particular proteins such as histones, ferritin, tubulin, and growth factors (see Chapter 7).

The polycistronic mRNA of prokaryotes usually has a 5' leader sequence starting with either pppA or pppG and there are untranslated intercistronic sequences as well as a 3'-untranslated region after the termination codon of the last cistron. The length of the leader sequence varies depending on the mRNA but is usually short (approximately 20–200 nucleotides). Exceptions are known where the initiation codon is located at or only a few nucleotides from the 5'-end of the mRNA. In prokaryotes, translation of nascent mRNA starts before transcription is terminated and the presence of ribosomes on the leader can alter the secondary structure of the nascent mRNA, thereby regulating the transcription of the rest of the operon by attenuation. Also in some cases leader sequences may be translated into short non-functional polypeptides.

Intercistronic sequences usually tend to be short, but may be as long as 400 nucleotides or absent altogether. There are also examples of overlapping cistrons where the last nucleotide of a UGA termination codon serves as the first residue of the 3'-proximal AUG initiation codon.

The function of the 3'-untranslated region of mRNA in protein synthesis is unclear; its main role may be in the termination of transcription by RNA polymerase but it may also be a determinant of mRNA stability and translational control (see Section 1 of Chapter 7).

Generally, eukaryotic mRNAs differ from prokaryotic messengers in two major features, namely a 5' cap structure and a 3' poly(A) tail. Caps consist of a 7-methylguanine nucleotide linked by a 5'-pyrophosphate bridge to the 5'-end of the mRNA, in which the ribose of the first two nucleotide residues may be methylated in the 2' position. The structure is known as cap 0 when neither ribose is methylated, cap 1 when the first ribose is methylated and cap 2 when both ribose residues are methylated (*Figure 4.3*). A few eukaryotic messengers may lack the cap structure. For example, the 5'-end of polysomal poliovirus RNA starts with pUp (2). The mRNAs of mitochondria and chloroplasts also lack a cap structure, possibly reflecting their prokaryotic origin.

A poly(A) tail of variable length (usually between 25 and 250 nucleotides) is present at the 3'-end of most but not all eukaryotic mRNAs. A notable exception is the family of histone messengers which, as a rule, lack poly(A). Apart from this example there seem to be few if any qualitative differences between

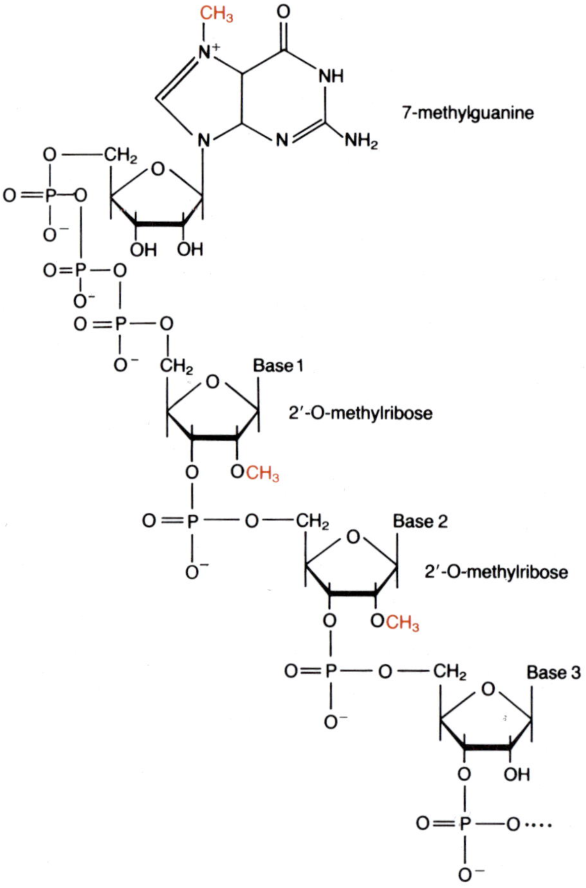

Figure 4.3. Cap structure of eukaryotic messenger RNAs. All caps contain 7-methylguanylate which is linked through two pyrophosphate bonds to the 5'-terminal nucleotide of the mRNA. Cap 0 structures lack *O*-methylated ribose, cap 1 structures contain 2'-*O*-methyl ribose in the first nucleotide and cap 2 structures contain 2'-*O*-methyl ribose in both the first and second nucleotide, as shown in the figure.

poly(A)$^+$ and poly(A)$^-$ messengers. It should be noted that the latter group is defined only operationally by its inability to bind to oligo(dT) cellulose or similar affinity columns, and may therefore include mRNAs containing short stretches of A residues which may be insufficient to allow binding. Poly(A) is also present in some prokaryotic messengers such as the mRNA for the lipoprotein of the *E.coli* outer membrane, which contains a short poly(A) sequence of $10-15$ residues at the 3′-end (3).

The poly(A) tail is not required for translation but increases the stability of some messengers such as globin mRNA. *In vivo*, multiple copies of a specific protein (M_r 76 000) are bound to the poly(A) segment; this may prevent degradation by nucleases. A number of other proteins are also present in close association with eukaryotic mRNAs but their location and function have not been elucidated in any detail (4).

Since the genetic code (see Chapter 1) is comma-less, accurate translation of mRNA requires the selection of the appropriate reading frame by precise interactions between the AUG initiation codon (or, in rare cases, alternative initiation codons), charged initiator tRNA and ribosomal subunits. This process takes place in a number of steps requiring the participation of several initiation factors, as discussed in Sections 2.3–2.5.

2.2. Initiator tRNA

All protein synthesizing systems are equipped with a special initiator tRNA, termed Met-tRNA$_f$ or Met-tRNA$_i$ when charged with the amino acid, which recognizes both the initiation codon of mRNA and the ribosomal P-site; it is the only tRNA that can enter this site directly, thus placing methionyl-tRNA in the correct position for making the first peptide bond. The initiation codon is usually AUG, but occasionally GUG is used and recently, initiation of translation of a viral mRNA has been observed at the ACG triplet in mammalian cells (5). In mammalian mitochondria AUA serves as the initiation codon.

In prokaryotes, the methionyl-tRNA$_f$ is formylated by a transformylase using N^{10}-formyltetrahydrofolate as the formyl group donor. Formylation of Met-tRNA$_f$ also occurs in mitochondria of eukaryotes but not in the cytosol where the initiator tRNA remains unformylated. Interestingly, cytosolic Met-tRNA$_f$ from eukaryotes can be formylated by the bacterial system and formylated Met-tRNA$_f$ is able to function in cell-free eukaryotic systems. Thus, these differences in the formylation of initiator tRNA appear to be due to the absence of the transformylase from the cytosol of eukaryotes. Methionyl-tRNA$_m$, which inserts the amino acid into internal positions of the growing peptide chain, is not formylatable.

2.3. Pre-initiation complexes

The sequence of events leading to the formation of the initiation complex from mRNA, charged initiator tRNA and ribosomal subunits involves a number of pre-initiation complexes as intermediates (*Figure 4.4*). In outline, prokaryotic and eukaryotic systems are similar, but there are some differences, particularly as

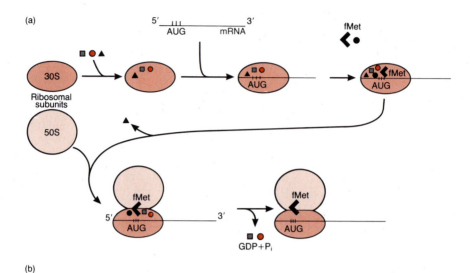

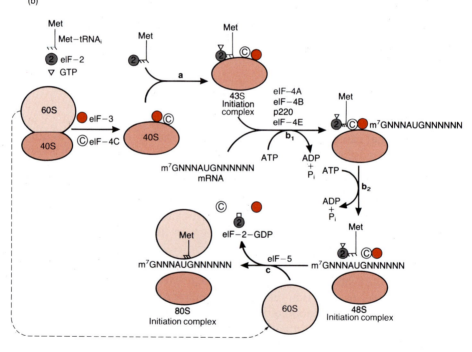

Figure 4.4. Schematic diagram illustrating the formation of initiation complexes. **(a)** Prokaryotic initiation. The following symbols are used for initiation factors and other components involved in the formation of the 70S initiation complex: □ IF-1, ○ IF-2, △ IF-3, ◀ fmet – tRNA$_f$, ● GTP. **(b)** Eukaryotic initiation complexes in mammalian protein biosynthesis (Rhoads 1988). The process may be divided into three stages: (a) formation of a 43S pre-initiation complex; (b) binding of messenger RNA with formation of a 48S pre-initiation complex; and (c) synthesis of the 80S initiation complex containing the initiator tRNA in the correct position for peptide bond formation.

regards the complexity of the initiation factors and details of the mechanisms involved.

2.3.1. Prokaryotic systems

Ribosomal subunits are generated by dissociation of ribosomes following translation of the mRNA. At physiological concentrations of magnesium ions, the rate of dissociation is slow and so dissociation is facilitated by the combined action of the initiation factors IF-1 and IF-3 (see *Table 4.1*); IF-1 increases the rate of dissociation and IF-3 acts as an anti-association factor when bound to the 30S ribosomal subunit, thereby displacing the equilibrium in favour of subunit formation. Initiation factor IF-2 is also able to bind to the 30S subunit and this association is stabilized by IF-1 and GTP, the latter acting as a steric effector without being hydrolysed at this stage. IF-2 plays a central role in binding fMet-tRNA$_f$ to the 30S pre-initiation complex by specific recognition of the *N*-formylmethionine residue attached to the tRNA, thus restricting this interaction to charged initiator tRNA. All three factors bind to the 30S ribosomal subunit near the 3'-end of the 16S ribosomal RNA at adjacent sites that are located at the interface between the small and large ribosomal subunits (see *Figure 3.4*).

   In the next step, the initiator tRNA and mRNA associate with the $30S \cdot IF\text{-}1 \cdot IF\text{-}2 \cdot IF\text{-}3$ complex with release of IF-3. There is evidence from *in vitro* experiments that the binding of mRNA precedes that of the initiator tRNA. The mechanism involved in the specific binding of mRNA will be considered later (Section 2.4).

2.3.2. Eukaryotic systems

The properties and functions of the eukaryotic initiation factors are listed in *Table 4.2*. The dissociation of cytosolic 80S ribosomes is facilitated by a complex initiation factor, eIF-3 ($M_r \approx 5-700\,000$), consisting of $9-11$ polypeptide chains, that binds to the small ribosomal subunit (40S) and prevents its re-association with the 60S subunits to form 80S ribosomes. Thus, this factor has anti-association activity, but low molecular weight proteins with similar activity have also been reported. Also, a small protein, eIF-4C, (M_r 20 000) seems to function as an accessory factor to eIF-3 in the formation of a 43S ribosomal

Table 4.1. Properties of prokaryotic initiation factors from *E.coli*

Factor	M_r (kDa)	Properties and function
IF-1	9	Stimulates rate of ribosome dissociation and activity of IF-2
IF-2	100	Binds fMet-tRNA$_f$ to the ribosomal P-site by a GTP-requiring reaction
IF-3	22	Binds natural mRNAs to small ribosomal subunit probably by facilitating base pairing between the untranslated leader sequence and the 3'-end of 16S rRNA

Table 4.2. Eukaryotic initiation factors

Initiation factor	Synonym	M_r (kDa) of factor or subunits	Properties and function
eIF-1		15	Stabilizes initiation complexes
eIF-2		$\alpha(38^*)$, $\beta(35-50^*)$, $\gamma(55)$	GTP-dependent binding of Met-tRNA$_f$ to small ribosomal subunit
eIF-2B	GEF	27,37,52,67,85*	Conversion of eIF-2·GDP into eIF-2·GTP
eIF-3		9–11 subunits of M_r 24–170*	Associates with 40S subunit to maintain dissociation; binds mRNA to 43S preinitiation complex
eIF-4A	50 kDa component of CBP-II	50	ATP-dependent unwinding of secondary structure of mRNA 5′ region. Stimulates translation of exogenous mRNA in cell-free system
eIF-4B		80*	mRNA binding; stimulates cell-free translation, ATPase activity of eIF-4A and eIF-4F, AUG recognition and recycling of eIF-4F
eIF-4C		17	Ribosome dissociation; 60S subunit joining
eIF-4D		17	Formation of first peptide bond
eIF-4E	CBP-I; 24 kDa CBP	24–28*	Binds mRNA cap structure
eIF-4F	CBP-II; cap-binding protein complex	24 (CBP-I)*, 50 (eIF-4A), 200*	ATPase; unwinds mRNA secondary structure; stimulates cell-free translation
eIF-5		60*	GTPase; release of eIF-2 and eIF-3 from pre-initiation complex to allow joining of 60S subunit
eIF-6		24	Anti-association activity; binds to 60S ribosomal subunit

*Denotes subunit can be phosphorylated *in vivo*.
Abbreviations: GEF, guanine nucleotide exchange factor; CBP, cap-binding protein.

pre-initiation complex and another protein factor, eIF-6 (M_r 24 000) prevents re-association by binding to the large (60S) ribosomal subunit.

Initiation factor eIF-2 gives a stable binary complex with GTP which binds the initiator tRNA, Met-tRNA$_f$, forming a ternary complex. Interaction of this ternary complex with the 40S ribosomal subunit containing bound initiation factors eIF-3 and eIF-4C gives rise to the 43S pre-initiation complex, which is competent to bind mRNA in the presence of three further initiation factors, eIF-4A, eIF-4B, and eIF-4F, together with ATP.

2.4. Binding of mRNA to the small ribosomal subunit

The mRNA binds to the small ribosomal subunit immediately before formation of the final initiation complex and requires the correct positioning of the initiation codon. In the case of bacterial and bacteriophage messengers the molecular recognition mechanism proposed by Shine and Dalgarno (6) involves base pairing between short nucleotide sequences, most often CUCC, near the 3'-end of the 16S rRNA and a complementary region, usually consisting of 3 – 9 bases on the 5'-side (upstream) of the mRNA initiation codon, which has been found to be present in nearly all of more than 150 bacterial and bacteriophage messengers. Studies with mutants and mRNA fragments indicate that, in addition to the Shine – Dalgarno interaction, outlying upstream sequences in the leader sequence may also provide recognition signals between mRNAs and ribosomes, possibly by ensuring that the Shine – Dalgarno sequence is in an appropriate conformation.

The Shine – Dalgarno mechanism is also found in chloroplast protein synthesis as judged from limited sequence analysis of the 16S rRNA and mRNAs, but apparently not in mammalian mitochondrial systems where the initiator codon occurs either directly at, or only a few nucleotides downstream from, the 5'-end of mRNA, which excludes the possibility of mRNA – rRNA base pairing in this region.

The binding of cytosolic eukaryotic mRNAs to the small ribosomal subunit probably does not involve base pairing with the 18S rRNA since no uninterrupted sequences of the Shine – Dalgarno type have been found. Instead, a 'scanning model' has been proposed in which the pre-initiation complex, composed of the 40S ribosomal subunit, Met-tRNA$_f$, and associated initiation factors, binds at or near the 5'-cap of the mRNA and slides along the messenger until it encounters the first AUG triplet, at which point the 60S ribosomal subunit joins to give rise to the 80S initiation complex (*Figure 4.4b*). Recognition of the cap is facilitated by cap-binding proteins, which mediate an ATP-dependent melting of the mRNA secondary structure at the 5'-terminal region to allow the mRNA to thread through a channel in the neck of the 40S subunit. The cap structure is required for efficient binding and translation even in cases where the initiating AUG codon occurs hundreds of nucleotides downstream. As a rule, scanning by the 40S subunit stalls at the first AUG codon, which is recognized mainly by interaction with the anticodon of the Met-tRNA$_f$. However, this recognition also depends in some way on eIF-2 and may be modulated as a result of deviation of the mRNA structure from the consensus sequence GCCGCC$_G^A$CCAUGG. Sequence context may also account for the rare cases where initiation occurs downstream of the 5'-proximal AUG codon. Where this sequence context is unfavourable, initiation becomes inefficient and hence most 40S subunits will tend to initiate further along the mRNA at another AUG triplet in a more favourable context. This model also explains rare cases where initiation is not restricted to one particular AUG codon and translation of a single mRNA gives rise to two proteins.

2.5. The final step in initiation: joining of the large ribosomal subunit

The last event in the initiation of protein synthesis involves the joining of the large ribosomal subunit to the pre-initiation complex. In the prokaryotic system, association of the 50S subunit with the 30S pre-initiation complex takes place with hydrolysis of GTP by the GTPase activity of IF-2 and the release of IF-1, IF-2, GDP, and P_i. GTP hydrolysis is essential for the release of IF-2 from the initiation complex, which is a prerequisite for allowing the fMet-tRNA$_f$ to engage in the formation of the first peptide bond. In eukaryotic protein synthesis, the 80S initiation complex is formed by joining the 60S ribosomal subunit to the 48S pre-initiation complex consisting of the 40S ribosomal subunit, eIF-2, eIF-3, GTP, Met-tRNA$_f$, mRNA, and possibly eIF-4C. This coupling reaction requires an additional factor, eIF-5, which mediates the hydrolysis of GTP to GDP with release of eIF-2·GDP, P_i, and eIF-3 from the 48S pre-initiation complex.

By this stage, all of the initiation factors have been released and are available for recycling, although the exact steps at which factors are released from intermediate complexes are not known in every case. There is thus an initiation factor cycle within the ribosome cycle and regulation of the activity of factors, particularly eIF-2, is an important control mechanism in translation (see Section 3.1 of Chapter 7).

In the initiation complex, location of the charged initiator tRNA in the P-site of the ribosome allows transfer of the methionine residue to the amino group of another aminoacyl-tRNA in the A-site by peptidyl transferase to form dipeptidyl-tRNA. Functional insertion of Met-tRNA$_f$ directly into the P-site can be demonstrated using the trinucleotide AUG as a synthetic mRNA and another trinucleotide, for example UUU, can be used to bind an acceptor aminoacyl-tRNA (in this case Phe-tRNA), leading to the synthesis of Met·Phe·tRNA. The antibiotic puromycin, which resembles the 3'-terminal region of Phe-tRNA in structure, can be used as an artificial acceptor in the peptidyl transferase reaction in the absence of mRNA, resulting in the formation of Met·puromycin.

3. Polypeptide chain synthesis: the elongation – translocation cycle

3.1. Peptide bond synthesis

The first peptide bond is formed when the aminoacyl-tRNA in the ribosomal A-site is converted into the corresponding methionyl-aminoacyl-tRNA by transfer of the methionyl (or *N*-formylmethionyl) residue from the charged initiator tRNA in the P-site. Any *N*-substituted aminoacyl-tRNA, such as peptidyl-tRNA or *N*-acetylaminoacyl-tRNA, can function in peptide bond synthesis as a donor in the P-site in place of the charged initiator tRNA. The reaction is catalysed by peptidyl transferase, which is an integral part of the large ribosomal subunit, as shown by the activity of isolated subunits. No soluble cofactors appear to be involved, but monovalent cations (K^+) at a concentration of 100 mM or more and divalent cations (Mg^{2+}) below 2 mM are required.

Efficient entry of aminoacyl-tRNA into the ribosomal A-site requires the participation of an elongation factor (termed EF-Tu in prokaryotes and EF-1$_\alpha$ (EF-1$_L$) in eukaryotes, see *Table 4.3*) and GTP. This elongation factor forms a ternary complex with GTP and all aminoacyl-tRNAs except the initiator tRNA. No ternary complex is formed with uncharged tRNA, thus ensuring that only appropriately charged tRNAs are efficiently bound in the A-site. A special elongation factor showing extensive homology with both EF-Tu and IF-2 is involved in the synthesis of selenoproteins (see Section 8.1 of Chapter 1) from selenocysteyl-tRNA^{UCA} in *E.coli* (7). In the EF-Tu-mediated aminoacyl-tRNA binding reaction there is no hydrolysis of GTP, but GTP must be hydrolysed before peptide bond synthesis can take place. The consequent delay between binding of the aminoacyl-tRNA to the A site and peptide bond formation enhances the fidelity of translation because the required stability of the tRNA/mRNA interaction is favoured by correct base pairing between the anticodon and codon, whereas any mismatched complexes will tend to dissociate before the peptide bond can be synthesized.

Table 4.3. Properties of elongation and termination factors

	M_r (kDa)	Properties and function
Elongation factors from *E.coli*		
EF-Tu	43	Amino terminal acetylserine; heat labile; binds aminoacyl-tRNA to the ribosomal A-site
EF-Ts	30	Regeneration of EFTu·GTP; heat stable
EF-G	77	GTP-dependent translocation of peptidyl-tRNA and its mRNA codon from the A-site to the P-site of the ribosome
Termination (release) factors from *E.coli*		
RF1	36	Requires UAA or UAG codons for hydrolysis of peptidyl-tRNA
RF2	38	Requires UAA or UGA codons for hydrolysis of peptidyl-tRNA
RF3	46	Enhances RF1 and RF2 activity
Eukaryotic elongation factors		
EF-1$_\alpha$ (EF-1$_L$ or eEF-Tu)	50–60	Analogous to EF-Tu
EF-1$_\beta$ (eEF-Ts)	30	Analogous to EF-Ts
EF-2	105	GTP-dependent translocation analogous to EF-G; contains essential SH groups and one residue of a post-translationally modified histidine called diphthamide
EF-3	125	GTPase and ATPase activity; function not fully defined
Termination factor from reticulocytes		
RF	110	Two 55 kDa subunits; binds to the ribosomal A-site by a GTP- and termination codon-dependent reaction; hydrolyses peptidyl-tRNA in the P-site

After GTP hydrolysis, the elongation factor is released from the ribosome as a complex with GDP. The mechanism of this elongation factor thus resembles that of initiation factor IF-2 in the binding of charged initiator tRNA to the small ribosomal subunit.

Following dissociation from the ribosome the EF-Tu·GDP complex interacts with another elongation factor, EF-Ts, with formation of an EF-Tu·EF-Ts heterodimer and release of GDP. Reaction of the heterodimer with GTP regenerates the EF-Tu·GTP complex required for binding aminoacyl-tRNA. In eukaryotes the sequence of events is similar with EF-1_α (M_r 50 000) corresponding to EF-Tu and EF-1_β (M_r 30 000) to EF-Ts.

Selection of the specific aminoacyl-tRNA to be bound at the ribosomal A-site is by base pairing between the relevant mRNA codon and the tRNA anticodon. Since this interaction involves only a triplet of bases and hence a maximum of 9 hydrogen bonds, it is intrinsically unstable at physiological temperatures and is probably stabilized by components of the ribosome to allow sufficient time for peptide bond synthesis to occur. Also, the codon – anticodon pairing must be monitored for fidelity in order to minimize errors in translation. There is genetic and biochemical evidence in *E.coli* for the involvement of one of the proteins of the small ribosomal subunit, S12, in ensuring the fidelity of normal translation and in the mistranslation which occurs in the presence of the antibiotic streptomycin due to incorrect codon – anticodon interactions.

3.2. Translocation

Immediately after synthesis of the first peptide bond, the ribosomal A-site contains dipeptidyl-tRNA whilst uncharged initiator tRNA remains in the P-site. Thus, both these sites are occupied, and to allow the next aminoacyl-tRNA to enter the A-site it is necessary to eject the uncharged tRNA and shift the dipeptidyl-tRNA from the A-site into the P-site. This translocation takes place as a concerted process involving movement of both mRNA and dipeptidyl-tRNA together into the P-site, leaving the A-site occupied by the next mRNA codon and free to accept the cognate aminoacyl-tRNA. According to the two-site model of the ribosome, ejection of the deacylated initiator tRNA occurs at the same time, but there is increasing evidence for a three-site model which postulates movement of the deacylated tRNA first to an E-(exit) site with subsequent ejection when the next aminoacyl-tRNA enters the A-site (*Figure 3.7*).

Translocation requires the participation of another elongation factor (termed EF-G in prokaryotes and EF-2 in eukaryotes) and GTP. It seems that when EF-G and GTP bind to the ribosome translocation occurs but that GTP hydrolysis is required only subsequently to release EF-G and GDP. The location of the EF-G binding site on the ribosome overlaps with that for EF-Tu and thus EF-G must be released before the EF-Tu·aminoacyl-tRNA·GTP complex can enter the A-site. Analogous reactions occur in eukaryotic systems.

There is little information about the details of the translocation mechanism. A continuous polyribonucleotide chain is not essential since translocation can occur with individual trinucleotides. It seems likely that movement of the mRNA is dependent on and tightly coupled to that of the tRNA with the binding sites for the tRNA providing the precision for movement by exactly one codon. Presumably, binding of EF-G and GTP after release of EF-Tu·GDP following peptide bond synthesis induces a conformational change in the ribosome which leads to translocation.

After the first translocation the ribosomal P-site is occupied by dipeptidyl-tRNA and the vacant A-site contains the third mRNA codon. Entry of the next aminoacyl-tRNA, selected as before by the codon–anticodon interaction, into the A-site enables peptide bond synthesis to continue and in this way repeated operation of the elongation–translocation cycle gives rise to a stepwise elongation of the nascent polypeptide chain, each complete cycle elongating the chain by one amino acid residue and moving the mRNA by one codon in the 5' to 3' direction. When the end of the coding sequence is reached and one of the termination (or stop) codons has entered the A-site, translation stops and the completed polypeptide chain is released.

3.3. Termination

The presence of one of the three termination codons, UAA, UAG, or UGA, in the A-site results in the binding to the ribosome of a release factor instead of an aminoacyl-tRNA. In prokaryotes two release factors have been identified, one (RF1) recognizing UAA or UAG, the other (RF2) functioning with UAA or UGA. Ribosomal binding and release of RF1 and RF2 are stimulated by a third factor, RF3 or S, which interacts with GTP and GDP. In eukaryotic cells such as reticulocytes, one release factor (RF) has been found to function with all three termination codons and the binding of this factor to ribosomes is stimulated by GTP but not GDP. Although the details are not entirely clear, GTP hydrolysis appears to be required for the release of the finished polypeptide chain by cleavage of the peptidyl-tRNA bond and completion of the termination process leading to dissociation of the release factor from the ribosome.

Thus, at the end of the ribosome cycle the coding sequence of mRNA has been translated to produce a polypeptide chain and all the components which have been involved in protein synthesis are available for re-use in another round of the cycle.

4. Further reading

Bielka,H. (1985) Properties and spatial arrangement of components in preinitiation complexes of eukaryotic protein synthesis. *Prog. Nucleic Acid Res. Mol. Biol.*, **32**, 267–89.

Kozak,M. (1983) Comparison of initiation of protein synthesis in procaryotes, eucaryotes, and organelles. *Microbiol. Rev.*, **47**, 1–45.

Kozak,M. (1989) The scanning model for translation: An update. *J. Cell Biol.*, **108**, 229–41.

Moldave,K. (1985) Eukaryotic protein synthesis. *Annu. Rev. Biochem.*, **54**, 1109–49.

Pain,V.M. (1986) Initiation of protein synthesis in mammalian cells. *Biochem. J.*, **235**, 625 – 37.

Rhoads,R.E. (1988) Cap recognition and the entry of mRNA into the protein synthesis initiation cycle. *Trends Biochem. Sci.*, **13**, 52 – 6.

Sonenberg,N. (1988) Cap-binding proteins of eukaryotic messenger RNA: Functions in initiation and control of translation. *Prog. Nucleic Acid Res. Mol. Biol.*, **35**, 173 – 207.

5. References

1. Nelson,E.M. and Winkler,M.M. (1987) *J. Biol. Chem.*, **262**, 11501 – 6.
2. Pettersson,R.F., Flanegan,J.B., Rose,J.K., and Baltimore,D. (1977) *Nature*, **268**, 270 – 2.
3. Taljanidisz,J., Karnik,P., and Sarkar,N. (1987) *J. Mol. Biol.*, **193**, 507 – 15.
4. Brawerman,G. (1987) *Cell*, **48**, 5 – 6.
5. Curran,J. and Kolakofsky,D. (1988) *EMBO J.*, **7**, 245 – 51.
6. Shine,J. and Dalgarno,L. (1974) *Proc. Natl. Acad. Sci. USA*, **71**, 1342 – 6.
7. Forchhammer,K., Leinfelder,W., and Böck,A. (1989) *Nature*, **342**, 453 – 6.

5

Protein folding and modifications of polypeptide chains in protein biosynthesis

1. Introduction

The translation mechanism discussed in the previous chapters gives rise to polypeptide chains with genetically-defined linear sequences containing the 20 common amino acids. Conversion of these primary products into the corresponding native proteins, which often exhibit major structural differences, involves specific changes such as removal of some amino acid residues from terminal or internal positions, the alteration of the side-chains of certain amino acids (*Table 5.1*) and folding of the polypeptide chains into the appropriate secondary and tertiary structure.

Protein folding may start as soon as the nascent polypeptide chain emerges from the ribosome and modifications to the primary structure may occur before, during or after folding of the polypeptide chain. Such modifications are said to be co-translational when the structural changes occur during elongation of the nascent polypeptide chains or post-translational if the completed polypeptide is modified after termination of peptide bond synthesis. The nature of the actual modification depends on the presence of both a specific amino acid sequence in the polypeptide and the appropriate modifying enzyme(s) in the relevant cell compartment(s). This combination of polypeptide synthesis and modification provides a versatile two-stage process which greatly enhances the range of protein structures that can be synthesized by the cell. As will become apparent in the following chapter, some modifications are directly related to the transfer of proteins to specific locations within the cell or to secretion of proteins. Moreover, modifications may be required for the formation of the correct secondary and tertiary structures of some native proteins, although in other cases newly synthesized polypeptides appear to acquire secondary and tertiary structure spontaneously.

Table 5.1. Modifications of amino acid residues in protein biosynthesis

Amino acid residue	Modifications
Alanine	Amino terminal methylation
Arginine	ADP – ribosylation; amino terminal methylation
Asparagine	ADP – ribosylation; glycosylation; amino terminal methylation; β-hydroxylation
Aspartate	Amide linkage to ethanolamine in GPI anchor; β-hydroxylation
Cysteine	Disulphide bond formation; fatty acylation
Glutamate	γ-Carboxylation; methylation
Glutamine	Cross-linkage to amino group of lysine; amino terminal methylation; internal cyclization to amino terminal pyroglutamic acid
Glycine	Conversion to carboxy terminal amide; amino terminal myristoylation
Histidine	Formation of diphthamide followed by ADP – ribosylation; amino terminal methylation
Lysine	Hydroxylation followed by glycosylation of 5-hydroxylysine; formation of cross-links; acetylation
Methionine	Deformylation of amino terminal formyl group; amino terminal methylation
Phenylalanine	Amino terminal methylation
Proline	Hydroxylation to 3- or 4-hydroxyproline; amino terminal methylation
Serine	Phosphorylation; glycosylation; fatty acylation; selenocysteine formation at tRNA level
Threonine	Phosphorylation; glycosylation; fatty acylation
Tyrosine	Phosphorylation; turnover of carboxy terminal residue in mammalian α-tubulin

2.   Protein folding

All proteins have secondary and tertiary structure which arise by folding of the linear polypeptide chain into a conformation that is stabilized by hydrogen bonds, van der Waal's forces, hydrophobic interactions, disulphide bonds and electrostatic interactions (see *Figure 1.1*).

2.1. Renaturation of proteins by folding in vitro

The classical experiments by Anfinsen showed that denatured pancreatic ribonuclease A is able to refold into the active enzyme under the appropriate conditions and similar results have been obtained with a number of other enzymes. In such cases, the primary amino acid sequence evidently contains all the information required for the acquisition of the correct protein conformation. Other proteins, however, require the participation of auxiliary proteins, termed molecular chaperones, in the folding process. For example, the reconstitution *in vitro* of active dimeric ribulose bisphosphate carboxylase from the unfolded polypeptides requires Mg·ATP and two chaperones, chaperonin-60 and chaperonin-10 (1). The chaperones are ubiquitous proteins which are known to participate together with ATP in the post-translational folding and assembly of various proteins in many different cells (2) (see also Sections 2 and 3 of Chapter 6).

The final three-dimensional structure of a protein is thermodynamically determined but the specific pathways involved in refolding are controlled by kinetic factors. The rates of protein folding reactions vary considerably. Experimentally, some proteins have been shown to regain their native conformation very rapidly (in milliseconds to seconds) whereas others refold only slowly (in minutes to hours). Different folding rates can be detected not only with different proteins but also in the same protein where fast- and slow-folding intermediates may co-exist. It is thought that one reason for such differences may be the co-existence of non-native isomers with abnormal disulphide and proline peptide bonds. The occurrence of such structures would be minimized under physiological conditions by the action of protein disulphide isomerase and prolyl isomerase, respectively.

2.2. *Protein folding* in vivo

Although the *in vitro* experiments provide an insight into the mechanisms involved in refolding denatured proteins, this process may differ from the folding of nascent chains that are continuously elongating by the stepwise addition of amino acid residues at the carboxy terminus (see Chapter 4) at the rate of approximately seven peptide bonds per second at 37°C. This rate is intermediate between the kinetics of fast and slow refolding of denatured proteins. Nascent polypeptide chains emerging from the ribosome during protein synthesis are therefore likely to assume thermodynamically and kinetically favourable intermediary conformations as elongation proceeds. Eventually, completed chains will be released from ribosomes and subsequently adopt a final native conformation which may differ significantly from the intermediate structures.

The extent to which nascent polypeptide chains are able to fold into regions of secondary and tertiary structures similar to or identical with those present in the final protein will also depend on the length of the nascent polypeptide outside the ribosome, some 30 amino acid residues being located within the ribosome and therefore, presumably, unable to undergo folding involving distant amino acid sequences. In long nascent chains, much of the secondary and tertiary structure may be similar to that of the final protein and examples are known where the translation of the mRNA for an enzyme gives rise to nascent polypeptides with sufficient secondary and tertiary structure to exhibit enzyme activity.

In some cases, such as insulin biosynthesis (see Section 3.2), a precursor protein acquires the appropriate secondary and tertiary structure which becomes stabilized by disulphide bridges (Section 3.3.1) before conversion into the final protein.

3. Structural modifications

Structural modifications fall into four broad categories, namely:

(i) hydrolytic removal of the terminal formyl group and/or methionine as well as cleavage of internal peptide bonds with or without removal of some amino acid residues;

(ii) covalent alterations of the carbon chains of certain amino acid residues such as proline, lysine, and glutamic acid;

(iii) substitution of amino acid residues by sugars or acetyl, carboxyl, alkyl, phosphate, or sulphate groups. Certain modifications, for example phosphorylations and acetylations, are readily reversible;

(iv) In some cases, extra amino acid residues may be added post-translationally. These modifications are rare and will not be discussed further.

3.1. Deformylation of the amino terminus and removal of the amino terminal methionine residue

In bacteria such as *E. coli* the amino terminal amino acid residue of most proteins are methionine, serine, or alanine. Since protein synthesis is initiated by fMet-tRNA$_f$, all newly synthesized polypeptides initially contain *N*-formylmethionine at the amino terminus. After translation formyl groups, as well as most methione residues are removed by a peptide deformylase and an aminopeptidase; the latter enzyme acts preferentially on methionyl peptide bonds (3).

In eukaryotic protein biosynthesis, the initiating methionine, which is not formylated, is often present only transiently during elongation of the nascent peptide. For example, in the biosynthesis of globin chains, the amino terminal methionine is removed when a chain length of about 30 amino acid residues is reached.

3.2. Proteolytic cleavage of polypeptide chains

In some cases, shortening of newly synthesized polypeptide chains by proteolytic cleavage is very extensive. Thus, the biosynthesis of insulin involves the initial synthesis of a larger precursor, preproinsulin, which is converted into proinsulin

Figure 5.1. Biosynthesis of insulin. Preproinsulin (I) is synthesized by ribosomes bound to the endoplasmic reticulum and translocated into the lumen with stabilization of the tertiary structure by formation of disulphide bonds (II) and proteolytic cleavage of the amino terminal signal sequence (pre-peptide; positions −1 to −24) at *a* to produce proinsulin (III) with the correct tertiary structure. Proinsulin is packaged into secretory vesicles in the trans-Golgi network where conversion into insulin is initiated. The secretory granules contain a trypsin-like endopeptidase (*b*) which cleaves on the carboxyl side of pairs of basic amino acids linking the connecting peptide (C-peptide, hatched circles) to the A- and B-chains and an exopeptidase (*c*) with a similar specificity as carboxypeptidase. Fully processed insulin (IV) is the first detectable in clathrin-coated vesicles budding from the trans-Golgi network. Amino acid residues are indicated by circles and identified where relevant by standard single letter abbreviations.

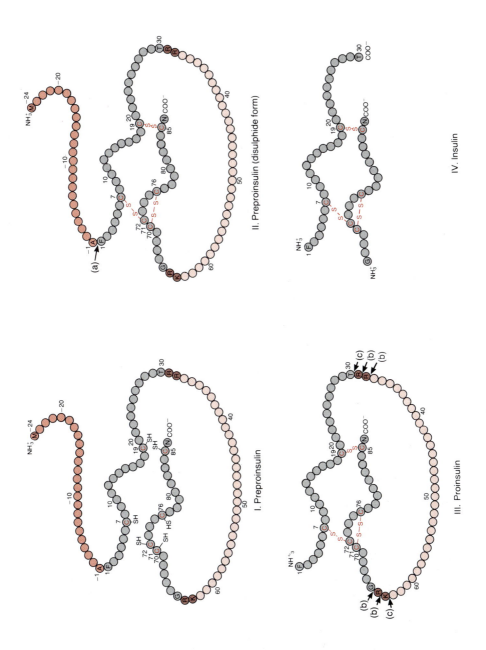

I. Preproinsulin

II. Preproinsulin (disulphide form)

III. Proinsulin

IV. Insulin

by loss of an amino terminal signal sequence followed by cleavage of proinsulin with release of four amino acids and a central connecting peptide containing 31 amino acid residues (*Figure 5.1*). This proteolytic maturation occurs during and after packaging of the precursor into clathrin-coated vesicles by the Golgi complex (4).

Equally extensive processing occurs in the biosynthesis of neuroendocrine peptides. Removal of the signal sequence from the initial translation product is followed by extensive further proteolytic cleavages either in the Golgi apparatus of the neuron before packaging into the synaptic vesicles or after packaging into the granules (5). In such cases, the primary translation product is sometimes known as a polyprotein since it functions as a precursor of several distinct proteins or peptides. Some viruses have also evolved complex proteolytic processing systems in which polyproteins are cleaved to give several structural and non-structural proteins (6). Thus, in these special situations, one gene can give rise to more than one protein product.

3.3. Modifications of amino acid side chains

3.3.1. Disulphide bond formation

Within the growing polypeptide chain the positions occupied by cysteine residues, like those of other amino acids, are governed by the genetic information in the mRNA. Some or all of the cysteine residues subsequently form intramolecular disulphide bonds which contribute to the stabilization of the final tertiary structure of the protein. The formation of disulphide bonds is a co-translational or post-translational process which is governed mainly by the three-dimensional structure arising from spontaneous folding of the new polypeptide chain. Correct disulphide bond formation is facilitated by the enzyme protein disulphide isomerase (PDI), which catalyses the rearrangement of disulphide bonds by an exchange reaction. In the synthesis of mammalian secretory and cell surface proteins, disulphide bond formation occurs in the lumen of the endoplasmic reticulum and the participation of PDI in the correct folding or assembly of disulphide-bonded proteins has been demonstrated by supplementation of a cell-free system containing PDI-deficient microsomes with the purified enzyme (7).

3.3.2. Hydroxylation of amino acid residues

The hydroxylation of certain proline and lysine residues is a well known post-translational modification in the biosynthesis of collagen. Proline residues within the sequences X – Pro – Gly, where X may be any amino acid except glycine, are hydroxylated mainly to 4-hydroxyproline, which is important for the stability of the collagen helix; 3-hydroxyproline also occurs in collagen but in smaller amounts. Some lysine residues are hydroxylated to 5-hydroxylysine, which then becomes glycosylated in a subsequent modification step. The enzymes involved in these hydroxylation reactions are located in the rough endoplasmic reticulum.

Although collagen is the major protein containing hydroxyproline and hydroxylysine, these modified amino acids also occur in other proteins. Thus, the core-specific lectin contains both hydroxyproline and hydroxylysine with some of the latter being O-glycosylated by glucosylgalactose groups (8).

Another hydroxyamino acid, β-hydroxyaspartate, occurs in certain proteins such as human factor IX, and together with β-hydroxyasparagine in the low density lipoprotein receptor as well as in bovine thrombomodulin (9).

3.3.3. Cross-linking of amino acid residues

The cross-linking of polypeptide chains in certain structural proteins such as collagen and elastin is a post-translational event which is vital for the stabilization of tissues and cellular matrices. Apart from disulphide bonds, a large number of different cross-bridges have been identified, and various amino acid residues are known to be involved in their formation. Well-known examples (*Figure 5.2*) include the oxidation of specific lysine and hydroxylysine residues in collagen to aldehydes, which condense with each other to form aldols or react with another lysine residue to give a secondary amine after reduction. Similarly, in the synthesis of elastin, three aldehyde groups arising by oxidation of lysine residues in different polypeptide chains condense with a lysine amino group with formation of a desmosine cross-link (10).

One of the most abundant cross-links involves the condensation of the amide group of a glutamine residue with the amino group of a lysine side-chain. This modification is important, for example, in the cross-linking of fibrin monomers during blood clotting. The reaction is catalysed by the calcium-dependent glutaminyl-peptide γ-glutamyl transferase known as glutaminase. The glutamine residue acts as the acyl donor whilst the primary amine group of a variety of compounds, including lysine and polyamines, can serve as the acceptor.

3.3.4. Carboxylation of glutamate residues

Prothrombin and several other plasma proteins involved in calcium-dependent interactions have been found to contain glutamate (Glu) residues that have been modified by post-translational carboxylation to γ-carboxyglutamate (Gla), which is able to chelate calcium. Proteins containing Gla residues are also present in brain and other tissues.

The conversion of glutamate to Gla residues is catalysed by a vitamin K-dependent carboxylase which is present in the endoplasmic reticulum. Studies of the substrate specifity of the enzyme using small peptides containing residues identical with those in positions 5 – 9 of prothrombin indicate that two consecutive glutamate residues are generally better substrates than a single glutamate. Sequences containing Asp – Asp are very poor substrates and in the case of an Asp – Glu peptide, only the glutamate residue was carboxylated.

The mechanism of the reaction is thought to involve the removal of a γ-hydrogen atom from the glutamate side chain by vitamin K, followed by carboxylation of the carbanion thus produced.

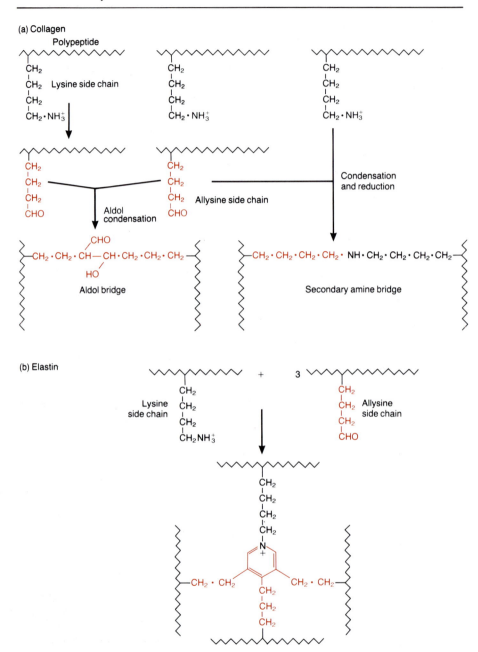

Figure 5.2. Formation of covalent inter-chain cross-links in collagen and elastin. (a) In collagen there are two types of cross-bridges which arise after oxidation of some lysine side chains to the aldehyde derivative (allysine), one involving an aldol condensation of two allysine residues as shown on the left, the other a reductive condensation between a lysine and an allysine residue (right). (b) In elastin, the combination of three allysine residues with one unmodified lysine side chain gives rise to a desmosine residue linking four polypeptide chains.

3.4. Substitution of amino acid residues

3.4.1. Glycosylation

In the biosynthesis of glycoproteins the introduction of mono- or oligo-saccharide units takes place by post-translational modification of polypeptides which become linked either through the amide group of asparagine or through the hydroxyl group of serine, threonine, or hydroxylysine residues. This complex process can give rise to a multiplicity of products arising both from differences in the oligosaccharides present at the same glycosylation site and also the diversity of glycosylation sites of any one polypeptide.

In the biosynthesis of asparagine-linked glycoproteins, the initial steps occur in the rough endoplasmic reticulum (RER) and involve firstly the stepwise assembly from nucleotide sugars of oligosaccharide precursors linked to a lipid carrier, dolichol phosphate, followed by the attachment of the oligosaccharide groups to the amide group of specific asparagine residues in accessible positions of the nascent acceptor polypeptide by oligosaccharyl transferases. These residues are usually present in the sequence Asn – X – Ser/Thr, where X can be any amino acid except proline or aspartic acid. Following transfer to the polypeptide, the oligosaccharide residues are processed, initially in the lumen of the RER by sequential removal of glucose residues by glucosidases, as well as of at least one α-1,2-linked mannose. Further modifications occur subsequently in the Golgi apparatus with the addition mainly of N-acetylglucosamine, fucose and sialic acid residues (*Figure 5.3*).

The biosynthesis of glycoproteins in which the sugars are linked to the hydroxyl groups of serine and threonine, or hydroxylysine in the case of collagen, involves the addition of specific sugars by different glycosyl transferases using either nucleoside diphosphate or nucleoside monophosphate sugars. The enzymes have a high degree of specificity with respect to both the donor nucleotide sugar and the position to be glycosylated. Some glycosylation may be initiated in the lumen of the endoplasmic reticulum as the nascent protein emerges, but most of the carbohydrate is added post-translationally in the Golgi apparatus. Many of the known glycosyl transferases are located in the lumen of the Golgi and addition of O-linked sugars to secretory proteins takes place at this site shortly before secretion.

3.4.2. Covalent modification of proteins by lipids

In both prokaryotes and eukaryotes, many membrane proteins are embedded in the lipid bilayers. The structural integrity of the membranes is maintained by non-covalent interactions of the lipids either directly with domains in polypeptide chains rich in hydrophobic amino acid residues or with fatty acids linked covalently to the relevant proteins. In many cases, the attached lipid not only anchors the protein to the membrane but also has additional functions.

A novel modification of several proteins important in the control of cell growth involves the transfer of an isoprenoid chain from farnesyl pyrophosphate to the thiol group of a specific cysteine residue, which subsequently becomes the

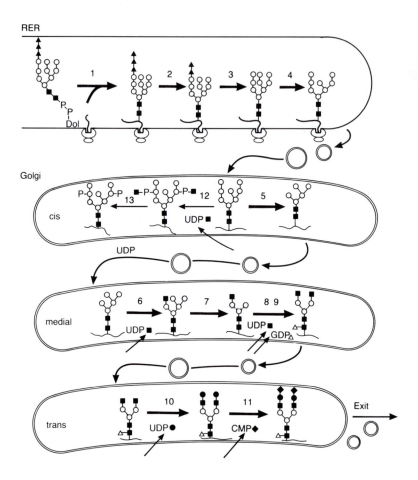

Figure 5.3. Structural modifications of oligosaccharides in glycoprotein synthesis (Kornfeld and Kornfeld, 1985). Oligosaccharide residues are attached to nascent proteins in the endoplasmic reticulum lumen and processing begins in this compartment but occurs mainly in the Golgi network. The following enzymes are involved: 1. oligosaccharyl transferase; 2. α-glucosidase I; 3. α-glucosidase II; 4. ER α-1,2-mannosidase; 5. Golgi α-mannosidase I; 6. N-acetylglucosaminyltransferase I; 7. Golgi α-mannosidase II; 8. N-acetylglucosaminyltransferase II; 9. fucosyl transferase; 10. galactosyltransferase; 11. sialyltransferase; 12. N-acetylglucosaminylphosphotransferase; 13. N-acetylglucosamine-1-phosphodiester α-N-acetylglucosaminidase. ■, N-acetylglucosamine; ○, mannose; ▲, glucose; △, fucose; ●, galactose; ◆, sialic acid.

carboxyl terminus of the modified polypeptide chain as a result of proteolytic cleavage (11).

Acylation of a variety of specific proteins by fatty acids is a ubiquitous modification in eukaryotic protein biosynthesis, occurring in yeast and plants as well as in both invertebrate and vertebrate animals. There are three main

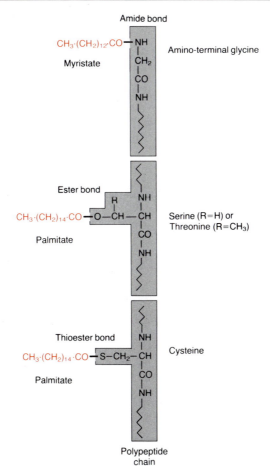

Figure 5.4. Covalent modification of polypeptides by fatty acids (adapted from R.J.A.Grand (1989) *Biochem J.*, **258**, 645–51, Figure 1).

types of protein–lipid linkages (*Figures 5.4* and *5.5*) involving different biosynthetic mechanisms, namely:

(1) post-translational esterification of fatty acids to cysteine, serine or threonine side chains;

(2) co-translational attachment of myristate to amino-terminal glycine residues; and

(3) post-translational linkage of glycosyl-phosphatidyl inositol membrane anchors *via* ethanolamine to an amino residue near the carboxyl terminus of a precursor polypeptide.

Ester-linked acylation. A wide variety of viral and cellular proteins, for example myelin proteolipid protein, viral glycoproteins, the human transferrin receptor, mucus glycoproteins, and the *ras* family of guanine nucleotide binding proteins, contain fatty acids covalently linked to the side chains of cysteine, serine, or threonine residues in polypeptides. Myristic, palmitic, stearic, and oleic acids

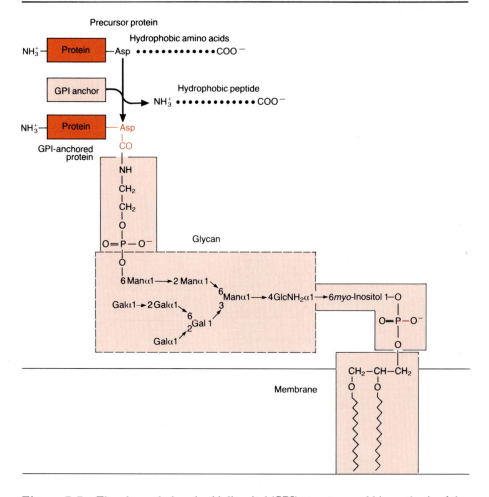

Figure 5.5. The glycosyl-phosphatidylinositol (GPI) structure and biosynthesis of the variant surface glycoprotein (VSG) of *Trypanosoma brucei* (adapted from M.A.J.Ferguson and A.F.Williams (1988) *Annu. Rev. Biochem.*, **57**, 285–320, Figure 1). The general structure and biosynthesis of other GPI-anchored proteins are similar but there are differences in the carboxyl terminal amino acid linked to the anchor and in the structure of the glycan. Also, the inositol moiety may be acylated by an additional fatty acid which becomes embedded in the membrane.

can all be esterified post-translationally to proteins in this way. The reactions involve fatty acyl CoA esters as donors and there is some specificity for particular fatty acids and proteins. In most cases, the precise physiological effect of acylation is unclear, but it is possible that it may serve to increase the concentration of these proteins in the membrane and thus enhance their function. *N-myristoylation*. In contrast to ester-linked acylation, protein *N*-myristoylation appears to take place either during or immediately after polypeptide chain elongation, as shown by the rapid inhibition of this acylation by inhibitors of

Table 5.2 Proteins with a glycosyl-phosphatidylinositol anchor (Low, 1989)

Hydrolytic enzymes	Alkaline phosphatase
	5'-Nucleotidase
	Acetylcholinesterase
	Trehalase
	Alkaline phosphodiesterase I
	p63 protease (*Leishmania*)
	Renal dipeptidase
	Merozoite protease (*Plasmodium*)
	Aminopeptidase P
	Lipoprotein lipase
Mammalian antigens	Thy-1
	RT-6
	Qa
	Ly-6
	Carcinoembryonic antigen
	Blast-1
	CD14
Protozoal antigens	Ssp-4 (*Trypanosoma*)
	Variant surface glycoprotein (*Trypanosoma*)
	Surface proteins (*Paramecium*)
	195 kDa antigen (*Plasmodium*)
Cell adhesion	LFA-3
	Heparan sulphate proteoglycan
	Neural cell adhesion molecule
	Contact site A (*Dictyostelium*)
	PH-20 (guinea pig sperm)
Miscellaneous	Decay-accelerating factor
	130 kDa hepatoma glycoprotein
	34 kDa placental growth factor
	Scrapie prion protein
	GP-2 (zymogen granule)
	Tegument protein (*Schistosoma*)
	FcIII receptor (human neutrophils)
	Oligodendrocyte-myelin protein
	Antigen 117 (*Dictyostelium*)
	125 kDa glycoprotein (*Saccharomyces*)
	Homologous restriction factor

protein synthesis. Although most *N*-myristoylated proteins are membrane-bound, some are found in the cytosol. The acylation reaction is catalysed by a transferase which uses myristoyl CoA as the acyl donor and prefers certain amino acid sequences in the substrate, but the specificity is not absolute and sequence context is also important. As is the case with ester-linked acylations, the function of *N*-linked myristoylation is uncertain but may be related to determining the intracellular destination of particular proteins as well as their redistribution within the cell in response to hormonal signals.

Inositol phospholipid membrane protein anchors. A number of membrane-associated proteins in a variety of organisms are attached to glycosyl-phosphatidyl inositol (GPI) anchors (*Table 5.2*). Biosynthetic labelling studies of the variant surface glycoprotein (VSG) (*Figure 5.5*) of the trypanosome, *Trypanosoma brucei*, indicate that linkage of the polypeptide to the anchor occurs within one minute of translation. In this and other examples of anchored proteins, removal of hydrophobic peptide sequences (predicted from DNA sequence determinations) present near the carboxyl terminus occurs rapidly, probably by a post-translational process, at the same time as attachment to the lipid. Thus, the original hydrophobic carboxyl-terminal sequence of the precursor polypeptide probably facilitates its initial retention in the lipid layer on the lumenal side of the endoplasmic reticulum for a sufficient time to allow transfer to the anchor structure. The enzymes involved in this process have not yet been characterized but the reaction is thought to be a transamidation in which the ethanolamine amino group of the glycophospholipid anchor forms an amide bond with an internal amino acid residue near the carboxyl terminus with cleavage of the carboxyl-terminal hydrophobic sequence at this point.

3.4.3. ADP – ribosylation

This family of post-translational modifications involves the transfer of one or several ADP – ribose groups from NAD^+ to an acceptor protein with the formation of either *N*- or *O*-glycosidic bonds between the ribose and an amino acid residue (*Figure 5.6*). The reactions are widespread and occur both in bacteria and eukaryotes. Mono-ADP – ribosylations link a single ADP – ribose unit to the side-chain nitrogen of an arginine, asparagine or diphthamide (a modified histidine) residue by an *N*-glycosidic bond. In mammalian cells, this reaction predominates, particularly in the extranuclear cell compartments. In addition to mammalian enzymes, several bacterial toxins such as diphtheria and cholera toxins have ADP – ribosyl transferase activity and modify certain cytosolic proteins in eukaryotes, for example elongation factor EF-2 and the G proteins of the adenylate cyclase complex, by mono-ADP – ribosylation. In contrast, poly-ADP – ribosylation occurs mainly in the nucleus and is initiated with formation of an *O*-glycosidic linkage between an ADP – ribose unit and the side-chain carboxyl group of a glutamate residue or the carboxyl group of a carboxyl-terminal lysine, further ADP – ribose units being added subsequently to produce a poly(ADP – ribose) chain. The principal substrates for poly-ADP – ribosylation are histones and various other nuclear proteins, including enzymes such as RNA polymerase α subunit and topoisomerase I.

3.4.4. Acetylation of proteins

Acetylation of amino terminal α-amino groups and of the ϵ-amino groups of internal lysine residues are common modifications of proteins. Indeed, in some eukaryotic cells such as Ehrlich ascites tumours, the majority of proteins may be *N*-acetylated (12). The precise step in polypeptide biosynthesis when amino terminal acetylation takes place has been investigated in only a few cases. In the biosynthesis of ovalbumin acetylation is a co-translational event which involves an acetylase bound to ribosomes and occurs soon after loss of the amino

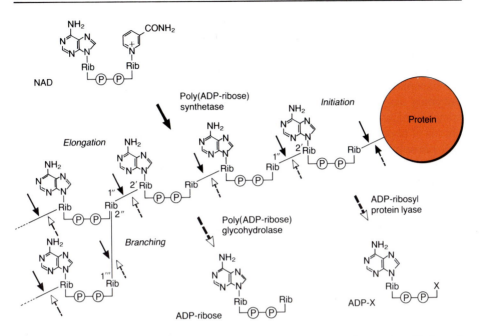

Figure 5.6. Biosynthesis and degradation of poly(ADP–ribosyl)-proteins (Ueda and Hayaishi, 1985). The solid arrows show steps in the biosynthetic pathway and open arrows indicate degradative reactions. The sugar X in the split product of ADP–ribose protein is 3-deoxy-D-*glycero*-pentos-2-ulose.

terminal methionine residue from the nascent chain containing about 44 amino acids compared with 20 amino acids when the methionine is removed (13). After acetylation, translation proceeds to completion when the acetylated polypeptide is released from the ribosome, but acetylation is not obligatory for completion of polypeptide chain synthesis. Some other proteins, such as actin and cat globin, may be acetylated post-translationally by a cytosolic enzyme. In all protein acetylations the donor is acetylCoA.

The acetylation of internal lysine residues of nuclear proteins such as histones differs in two major respects from the amino terminal acetylation reaction. First, the reaction is located in the nucleus, and secondly such acetyl groups are readily hydrolysed by a deacetylase. Thus, the extent of histone acetylation is the result of a dynamic equilibrium between acetylation and deacetylation. Increased acetylation levels of specific chromatin regions are associated with active gene expression.

3.4.5. Phosphorylation of proteins

Many proteins contain phosphate groups linked to the hydroxyl groups of serine, threonine, and tryosine residues in polypeptide chains. A few phosphoproteins such as casein and phosvitin are primarily of nutritional importance, but many enzymes, receptors, mediators, and regulatory factors are modified by reversible phosphorylation, which has important functions in cell growth and metabolic regulation. In fact, protein phosphorylation is the most common reversible post-translational modification and a multitude of kinases are known which catalyse the phosphorylation of specific serine, threonine or tyrosine residues of various enzymes and other proteins (14). Protein phosphorylation occurs post-translationally and phosphate groups turn over more rapidly than the polypeptide chains of phosphoproteins as a result of the combined action of protein kinases and phosphatases.

3.4.6. Protein methylation

The post-translational modification of proteins by methylation is relatively rare, but enzyme activities are known which use S-adenosylmethionine as the methyl donor to modify a number of different amino acid residues in a few specific proteins. Thus, bacterial membrane chemoreceptors are methylated by esterification of certain γ-glutamyl groups, possibly originally present as glutamine residues. The methyl group can be hydrolysed by an esterase and although this enzyme is not very active it is possible that reversible methylation may have a regulatory function. Other methylations of protein carboxyl groups are known but as yet not well characterized.

A second, different group of reactions involves the N-methylation of lysine, arginine, histidine, and glutamine residues. These modifications are responsible for the occurrence of methylated amino acids, such as dimethylarginine and 3-methylhistidine, in a variety of eukaryotic proteins including histones, flagellar proteins, myosin, actin, ribosomal proteins, fungal and plant cytochrome c, myelin basic protein, EF-Tu, EF-1α, calmodulin, heat shock proteins, and ferredoxin. Methylation of amino terminal amino acid residues is also known.

A third type of protein methylation reaction involving the esterification of a carboxyl-terminal carboxyl group by S-adenosylmethionine has been reported to modify the H-*ras* oncogene product in rat fibroblasts (15).

3.4.7. Introduction of carboxyl-terminal amide groups

Some proteins, including many peptide hormones, contain a carboxyl-terminal amide group, which is introduced by an oxidative modification of a carboxyl-terminal glycine residue in a precursor polypeptide (16). The reaction, which is catalysed by peptidylglycine hydroxylase, is thought to involve the initial hydroxylation of the glycine α-carbon atom, followed by a dismutation to the amide with loss of the glycine carbons as glyoxylate.

3.4.8. Protein tyrosine sulphation

The sulphation of tyrosine residues in proteins is a post-translational modification which is widespread amongst eukaryotic organisms. Although the proteins have different functions all of them are synthesized on the rough endoplasmic reticulum and are either secretory or membrane constituents. Sulphation of proteins involves the transfer of sulphate from 3'-phosphoadenosine 5'-phosphosulphate (PAPS) to specific tyrosine residues by tyrosylprotein transferase, which is an integral membrane protein of the trans-Golgi complex, that is the same compartment that is involved in the galactosylation and sialylation of asparagine-linked oligosaccharides in glycoprotein synthesis. The active site of the enzyme is oriented towards the lumen of the Golgi and PAPS is transported to this site from the cytosol by a specific transmembrane carrier.

In vivo, there is no significant turnover of protein tyrosine sulphate groups. Thus, protein sulphation appears to be irreversible and removal of sulphate occurs only after degradation of the sulphated polypeptides to free tyrosine sulphate.

4. Further reading

Goldenberg,D.P., Frieden,R.W., Haack,J.A., and Morrison,T.B. (1989) Mutational analysis of a protein-folding pathway. *Nature,* **338**, 127 – 32.

Huttner,W.B. (1987) Protein tyrosine sulfation. *Trends Biochem. Sci.,* **12**, 361 – 3.

Kim,P.S. and Baldwin,R.L. (1990) Intermediates in the folding reactions of small proteins. *Annu. Rev. Biochem.,* **59**, 631 – 60,

Kornfeld,R. and Kornfeld,S. (1985) Assembly of asparagine-linked oligosaccharides. *Annu. Rev. Biochem.,* **54**, 631 – 64.

Low,M.G. (1989) Glycosyl-phosphatidylinositol: a versatile anchor for cell surface proteins. *FASEB J.,* **3**, 1600 – 8.

Montelione,G.T. and Scheraga,H.A. (1989) Formation of local structures in protein folding. *Acc. Chem. Res.,* **22**, 70 – 6.

Ptitsyn,O.B. (1987) Protein folding: Hypotheses and experiments. *J. Prot. Chem.,* **6**, 273 – 93.

Roder,H., Elöve, G.A., and Englander,S.W. (1988) Structural characterization of folding intermediates in cytochrome *c* by H-exchange labelling and proton NMR. *Nature,* **335**, 700 – 4.

Schultz,A.M., Henderson,L.E., and Oroszlan,S. (1988) Fatty acylation of proteins. *Annu. Rev. Cell Biol.,* **4**, 611 – 47.

Towler,D.A., Gordon,J.I., Adams,S.P., and Glaser,L. (1988) The biology and enzymology of eukaryotic protein acylation. *Annu. Rev. Biochem.,* **57**, 69 – 99.

Ueda,K. and Hayaishi,O. (1985) ADP-Ribosylation. *Annu. Rev. Biochem.,* **54**, 73 – 100.

Yan,S.C.B., Grinnell,B.W., and Wold,F. (1989) Post-translational modifications of proteins: some problems left to solve. *Trends Biochem. Sci.,* **14**, 264 – 8.

Zappia,V., Galletti,P., Porta,R., and Wold,F. (eds) (1988) *Advances in post-translational modifications of proteins and aging.* Plenum Press, New York.

5. References

1. Goloubinoff,P., Christeller,J.T., Gatenby,A.A., and Lorimer,G.H. (1989) *Nature,* **342**, 884 – 9.
2. Ellis,R.J. and Hemmingsen,S.M. (1989) *Trends Biochem. Sci.,* **14**, 339 – 42.
3. Lucas-Lenard,J. and Lipmann,F. (1971) *Annu. Rev. Biochem.,* **40**, 409 – 48.
4. Orci,L., Ravazzola,M., Storch,M.-J., Anderson,R.G.W., Vassalli,J.-D., and Perrelet,A. (1987) *Cell,* **49**, 865 – 8.
5. Turner,A.J. (1986) *Essays Biochem.,* **22**, 69 – 119.
6. Strauss,E.G., Rice,C.M., and Strauss,C.H. (1984) *Virology,* **133**, 92 – 110.

7. Bulleid,N.J. and Freedman,R.B. (1988) *Nature,* **335**, 649–51.
8. Colley,K.J. and Baenziger,J.U. (1987) *J. Biol. Chem.,* **262**, 10290–5.
9. Stenflo,J. Öhlin,A.-K., Owen,W.G., and Schneider,W.J. (1988) *J. Biol. Chem.,* **263**, 21–4.
10. Gallop,P.M. and Paz,M.A. (1975) *Physiol. Rev.,* **55**, 418–87.
11. Goldstein,J.L. and Brown,M.S. (1990) *Nature,* **343**, 425–30.
12. Brown,J.L. and Roberts,W.K. (1976) *J. Biol. Chem.,* **251**, 1009–14.
13. Palmiter,R.D., Gagnon,J., and Walsh,K.A. (1978) *Proc. Natl. Acad. Sci. USA,* **75**, 94–8.
14. Hunter,T. (1987) *Cell,* **50**, 823–9.
15. Clarke,S., Vogel,J.P., Deschenes,R.J., and Stock,J. (1988) *Proc. Natl. Acad. Sci. USA,* **85**, 4643–7.
16. Bradbury,A.F. and Smyth,D.G. (1987) *Biosci. Rep.,* **7**, 907–16.

6

Targeting and translocation of proteins

1. Introduction

Many proteins are transported from their site of synthesis to remote locations. In eukaryotic cells most proteins are synthesized by cytosolic ribosomes and many enzymes involved in metabolism as well as specialized proteins such as haemoglobin and muscle proteins are retained in the cytosol, whereas others are translocated into cell organelles and membranes or exported from the cell. In addition, mitochondria and chloroplasts contain autonomous systems of protein synthesis comprising a relatively small number of less complex ribosomes which synthesize a few of the proteins specifically required in these organelles. Prokaryotes lack organelles and in these cells protein synthesis is not compartmentalized but, as in eukaryotes, some proteins are specifically retained in the cytosol whereas others are inserted into membranes or secreted by specific mechanisms (*Figure 6.1*).

The molecular basis for targeting proteins to their appropriate destinations has been elucidated in considerable detail. Both the export of proteins from cells and the translocation of proteins into mitochondria or other organelles in eukaryotes reside in specific signal sequences within the polypeptide chains themselves or their precursors. These signals direct either transfer from one subcellular compartment to another or retention at a particular location. Nascent polypeptides may contain more than one signal sequence which, for example, may specify translocation into, followed by retention within, say, the lumen of the endoplasmic reticulum.

There are two main alternative and mutually exclusive routes for protein translocation, involving the synthesis of polypeptides either by ribosomes bound to the endoplasmic reticulum or by free cytosolic ribosomes.

In the first pathway, protein synthesis gives rise to precursors containing an appropriate signal sequence, usually in the amino terminal region, which ensures the cotranslational insertion of the nascent protein into a specific channel in the

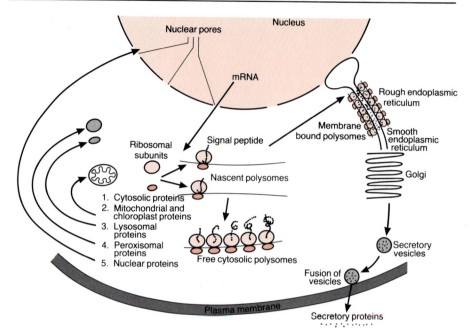

Figure 6.1. The main routes for targeting eukaryotic proteins. The biosynthesis of polypeptides is initiated by attachment of small and large ribosomal subunits either to newly synthesized mRNA that has been exported from the nucleus or to a free initiation site of polysomal mRNA during the operation of the ribosome cycle (Chapter 4). The pathways followed by the newly synthesized proteins to different destinations are indicated by arrows.

membrane followed by translocation into the lumen of the endoplasmic reticulum. After proteolytic cleavage of the signal peptide, the processed product may be retained in the lumen or further transported through the Golgi network into secretory vesicles or to other destinations depending on the particular protein. Retention of a protein in a given location appears to be governed by the presence of molecular signal sequences which prevent further transport.

The second major route also involves the synthesis of precursors but in this case transport to their destination takes place post-translationally, that is after completion of polypeptide chain synthesis in the cytoplasm. This mechanism operates particularly in the import of proteins from the cytosol into mitochondria and chloroplasts.

Interactions between newly synthesized proteins and their target in the relevant intracellular compartment may also be important, as for example in the case of nucleosome formation from histones and DNA in the nucleus. Such interactions are sometimes mediated not only by particular structural domains within the newly synthesized polypeptide chains but also indirectly by molecular chaperones which are proteins that facilitate the assembly of complexes without themselves being incorporated into the final structure (see also Section 2 of Chapter 5).

2. The co-translational translocation of polypeptides

Proteins which are to be translocated into the lumen of the endoplasmic reticulum (ER) are synthesized as precursors containing an amino terminal signal sequence or pre-sequence (as for example in preproinsulin). Typically, eukaryotic signal sequences comprise a core of hydrophobic residues within the first 20 – 30 amino acids at the amino terminal region of the newly synthesized polypeptide (*Table 6.1*). A few proteins such as ovalbumin contain internal sequences that are functionally equivalent to the amino terminal signal peptides, although they are not cleaved during translocation.

The signal sequences are recognized by a signal recognition particle (SRP) as soon as they emerge from the ribosome during elongation of the nascent polypeptide (*Figure 6.2*). The SRP also interacts with the ribosome carrying the nascent polypeptide chain, causing temporary arrest of translation. Next, the SRP – ribosome complex binds to an SRP receptor (also termed docking protein, DP) present in the ER, opening a channel by a GTP-requiring process. Another integral membrane protein, the signal sequence receptor (SSR), binds the signal sequence and facilitates entry of the nascent polypeptide chain into the translocation channel. At the same time, the SRP is released into the cytosol for recycling and translation resumes, resulting in translocation of the polypeptide chain into the ER lumen. Normally, the amino terminal signal sequence is removed before completion of polypeptide synthesis by a signal peptidase located in the endoplasmic reticulum. During translocation modifications of the polypeptide chain (Chapter 5) may occur, for example *N*-glycosylation of asparagine residues as the nascent polypeptide emerges into the lumen or fatty acylation of membrane proteins in the ER (1).

The SRP comprises a 7S RNA and six different polypeptide chains (M_r 9, 14, 19, 54, 68, and 72 kDa). The first two components function in translational arrest, the 54 kDa protein, which contains a GTP-binding domain (2), interacts with the signal sequence and the 68 kDa/72 kDa heterodimer binds to the docking protein. The docking protein has two subunits (a 70 kDa α and 30 kDa β polypeptide) and the SSR appears to be a hexamer containing two glycoproteins of 34 kDa and 23 kDa which may be involved in both translocation and processing of the nascent polypeptide by cleavage of the signal sequence.

3. Post-translational translocation

Many constituent proteins of organelles such as mitochondria (3), chloroplasts (4), lysosomes, and peroxisomes are synthesized by free ribosomes and released into the cytosol as precursors which are subsequently taken up by the organelle where processing to the final structure takes place. The relevant pathways are complex, as shown in detailed studies of yeast mitochondria (3). Thus, one route (A, *Figure 6.3*) has been dissected into four stages:

Table 6.1 Signal peptide sequences of eukaryotic secretory proteins

Protein	-29	-28	-27	-26	-25	-24	-23	-22	-21	-20	-19	-18	-17	-16	-15	-14	-13	-12	-11	-10	-9	-8	-7	-6	-5	-4	-3	-2	-1	1
Bovine preproalbumin												Met	Lys	Trp	Val	Thr	Phe	Ile	Ser	Leu	Leu	Leu	Leu	Phe	Ser	Ser	Ala	Tyr	Ser	Arg
Mouse immunoglobulin																														
L-chain MOPC-104E λ₁											Met	Ala	Trp	Ile	Ser	Leu	Ile	Leu	Ser	Leu	Leu	Ala	Leu	Ser	Ser	Gly	Ala	Ile	Ser	Gln
MOPC-41 κ								Met	Asp	Met	Arg	Ala	Pro	Ala	Gln	Ile	Phe	Gly	Phe	Leu	Leu	Leu	Leu	Phe	Pro	Gly	Thr	Arg	Cys	Asp
Human preproinsulin						Met	Ala	Leu	Trp	Met	Arg	Leu	Leu	Pro	Leu	Leu	Ala	Leu	Leu	Ala	Leu	Trp	Gly	Pro	Asp	Pro	Ala	Ala	Ala	Phe
Rat preproinsulin						Met	Ala	Leu	Trp	Met	Arg	Phe	Leu	Pro	Leu	Leu	Ala	Leu	Leu	Val	Leu	Trp	Glu	Pro	Lys	Pro	Ala	Gln	Ala	Phe
Rat preprolactin	Met	Asn	Ser	Gln	Val	Ser	Ala	Arg	Lys	Ala	Gly	Thr	Leu	Leu	Leu	Leu	Met	Met	Ser	Asn	Leu	Leu	Phe	Cys	Gln	Asn	Val	Gln	Thr	Leu
Human leukocyte interferon							Met	Ala	Ser	Pro	Phe	Ala	Leu	Leu	Met	Val	Leu	Val	Val	Leu	Ser	Cys	Lys	Ser	Ser	Cys	Ser	Leu	Gly	Cys
Human fibroblast interferon									Met	Thr	Asn	Lys	Cys	Leu	Leu	Gln	Ile	Ala	Leu	Leu	Leu	Cys	Phe	Ser	Thr	Thr	Ala	Leu	Ser	Met
Chicken lysozyme												Met	Arg	Ser	Leu	Leu	Ile	Leu	Val	Leu	Cys	Phe	Leu	Pro	Leu	Ala	Ala	Leu	Gly	Lys
Yeast invertase											Met	Leu	Leu	Gln	Ala	Phe	Leu	Phe	Leu	Leu	Ala	Gly	Phe	Ala	Ala	Lys	Ile	Ser	Ala	Ser

The amino acid residues of the presequences are given negative numbers starting at the amino terminus of the protein produced by cleavage of the signal peptide. Hydrophobic residues are coloured.

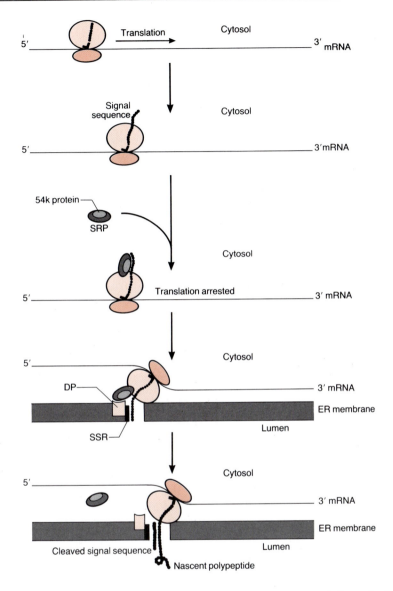

Figure 6.2. Co-translational transfer of proteins into the lumen of the endoplasmic reticulum. SRP, signal recognition particle; DP, docking protein (signal recognition particle receptor); SSR, signal sequence receptor; ER, endoplasmic reticulum.

(i) binding of the precursor protein to specific import sites, comprising a recently identified 42 kDa protein on the outer mitochondrial membrane (5), in the region where the outer and inner membranes are in close juxtaposition;

(ii) unfolding of the protein by association with an antifolding protein;

(iii) translocation across the inner membrane by a process requiring the

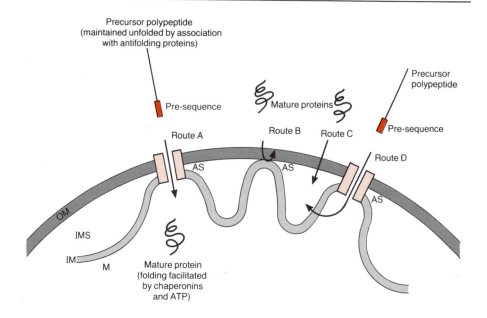

Figure 6.3. Protein translocation into mitochondria. The import of proteins from the cytosol into the four mitochondrial compartments by different pathways involves either modifications of precursors (routes A and D) or the translocation of mature-size proteins (routes B and C). AS, adhesion sites of inner and outer mitochondrial membranes; OM, outer membrane;   IMS, intermembrane space; IM, inner membrane; M, matrix.

    hydrolysis of ATP and an electrochemical gradient with a positive charge on the outside face of the membrane;

(iv) cleavage of amino terminal sequences, which are usually 10–20 amino acids in length, by a soluble metallo-protease present in the matrix followed by folding into the tertiary structure of the mature protein. This final stage may be facilitated by the participation of chaperones and ATP. Thus, one of the heat shock proteins, hsp60, has been shown to be essential for the assembly of proteins imported into yeast mitochondria (6).

    Another pathway (B) is involved in the import of proteins into the outer mitochondrial membrane, which takes place by the direct insertion of finished polypeptides after synthesis in the cytosol.

    Proteins that are found in the intermembrane space may follow two alternative routes from the cytosol. In one pathway (C) insertion of a mature-sized protein into the lipid bilayer of the outer membrane is followed by a molecular modification in the intermembrane space which facilitates translocation. Thus, cytochrome c crosses the outer membrane as the mature apoprotein, which then becomes covalently attached to haem in the intermembrane space, the haem protein being subsequently inserted into the outer face of the inner membrane. An alternative route (D) which is taken, for example, by cytochrome b_2, initially introduces a precursor into the matrix (route A). Following removal of the pre-sequence in two stages, first in the matrix and secondly at the inner membrane, the mature protein is released into the intermembrane compartment.

A number of post-translational modifications other than proteolytic cleavage of precursors may also be implicated in targeting proteins to particular locations. For example, precursors of some enzymes have been shown to require phosphorylation of certain mannose residues for import into lysosomes and fatty acylation may be important for the insertion of some proteins into plasma membranes (7).

4. Selective retention

Just as certain sequences within a polypeptide chain are important for recognition by receptors and subsequent stages of translocation through membrane channels, other domains can specify retention of specific proteins in particular cellular compartments. An interesting example of this mechanism is the retention of certain proteins within the lumen of the ER after cotranslational import. Whereas most proteins translocated into the lumen are processed further in the Golgi stack and eventually transferred to other locations within the cell or exported by exocytosis, a few are prevented from following this route and are retained in the lumen. The signal sequence for retention appears to be Lys – Asp – Glu – Leu at the carboxyl terminus, as shown by its presence in three ER proteins, protein disulphide isomerase and two glucose regulated proteins (grp78 and grp94), and the observation that addition of this sequence to lysozyme, which is normally secreted, leads to accumulation of the modified enzyme in the ER.

The uptake of specific proteins by the nucleus may involve two alternative mechanisms, one being dependent on specific interactions with other nuclear components, the other on nuclear location signals in the polypeptide chain. All nuclear proteins originate in the cytosol where they are synthesized on free ribosomes. Nuclear pores allow the rapid exchange of proteins of below 60 kDa and small proteins such as histones can therefore diffuse freely. Within the nucleus, histones associate into octamers which bind to DNA, giving rise to nucleosomes. *In vitro*, this assembly process requires the participation of a molecular chaperone, nucleoplasmin, an acidic nuclear protein which prevents the precipitation of histones by shielding their positive charges (8). The formation of this complex would be sufficient to ensure that histones are directed to and retained by the nucleus. Similarly, other small proteins with affinity for macromolecules such as DNA or nuclear RNA transcripts would also become localized in this organelle by such a mechanism, but proteins larger than 60 kDa only diffuse slowly and may therefore require specific sequences to facilitate entry into the nucleus (9). Using mutants of the nuclear protein SV40 large T antigen it has been shown that the nuclear location signal is distinct from the DNA binding domain (10). In this case, transport of the protein into the nucleus is independent of its association with DNA.

9. Further reading

Austen,B.M. and Westwood,O.M.R. (1991) *Protein targeting and secretion,* In Focus Series. Oxford University Press, Oxford.

Bergeron,J.J.M. (1988) Processing and targeting of proteins in the eucaryote. *Biochem. Cell Biol.*, **66**, 1253 – 7.

Bradshaw,R.A. (1989) Protein translocation and turnover in eukaryotic cells. *Trends Biochem Sci.*, **14**, 276 – 9.

Campbell,P.N. (1989) The targeting of proteins in eukaryotic cells. *Biochem. Education*, **17**, 114 – 21.

Eilers,M. and Schatz,G. (1988) Protein unfolding and the energetics of protein translocation across biological membranes. *Cell*, **52**, 481 – 3.

Ellis,R.J. and Hemmingsen,S.M. (1989) Molecular chaperones: proteins essential for the biogenesis of some macromolecular structures. *Trends Biochem. Sci.*, **14**, 339 – 42.

Lee,C. and Beckwith,J. (1986) Cotranslational and posttranslational protein translocation in prokaryotic systems. *Annu. Rev. Cell Biol.*, **2**, 315 – 36.

Lodish,H.F. (1988) Transport of secretory and membrane glycoproteins from the rough endoplasmic reticulum to the Golgi. *J. Biol. Chem.*, **263**, 2107 – 10.

Pelham,H.R.B. (1989) Control of protein exit from the endoplasmic reticulum. *Annu. Rev. Cell Biol.*, **5**, 1 – 23.

Pfanner,N. and Neupert,W. (1990) The mitochondrial protein import apparatus. *Annu. Rev. Biochem.*, **59**, 331 – 53.

Pfeffer,S.R. and Rothman,J.E. (1987) Biosynthetic protein transport and sorting by the endoplasmic reticulum and Golgi. *Annu. Rev. Biochem.*, **56**, 829 – 52.

10. References

1. Rose,J.K. and Doms,R.W. (1988) *Annu. Rev. Cell Biol.*, **4**, 257 – 88.
2. Römisch,K., Webb,J., Herz,J., Prehn,S., Frank,R., Vingron,M., and Dobberstein,B. (1989) *Nature*, **340**, 478 – 82.
3. Attardi,G. and Schatz,G. (1988) *Annu. Rev. Cell Biol.*, **4**, 289 – 333.
4. Smeekens,S., Weisbeck,P., and Robinson,C. (1990) *Trends Biochem. Sci.*, **15**, 73 – 6.
5. Vestweber,D., Brunner,J.,Baker,A., and Schatz,G. (1989) *Nature*, **341**, 205 – 9.
6. Cheng,M.Y., Hartl,F.-U., Martin,J., Pollock,R.A., Kalousek,F., Neupert,W., Hallberg,E.M., Hallberg,R.L., and Horwich,A.L. (1989) *Nature*, **337**, 620 – 5.
7. Burn,P. (1988) *Trends Biochem. Sci.*, **13**, 79 – 83.
8. Laskey,R.A., Honda,B.M., Mills,A.D., and Finch,J.T. (1978) *Nature*, **275**, 416 – 20.
9. Dingwall,C. and Laskey,R.A. (1986) *Annu. Rev. Cell Biol.*, **2**, 367 – 90.
10. Roberts,B. (1989) *Biochim. Biophys. Acta*, **1008**, 263 – 80.

7

Translational control of gene expression

1. Introduction

The capacity of a cell to synthesize a specific protein in the required amount at a particular time depends both on the availability of the relevant mRNA at the site of protein synthesis and on the efficiency with which it is translated. The steady state level of mRNA is determined by its rate of synthesis and degradation. In both prokaryotes and eukaryotes, the control of transcription is of major importance. In eukaryotes, factors involved in the processing of the initial transcript and its transport from the nucleus may also contribute to the regulation of gene expression but there is at present little information about their significance. The control of transcription is outside the scope of this book and this chapter will focus mainly on translational control mechanisms, that is factors which enhance or repress the translation of mRNA. Changes in the stability of translatable mRNA will also be considered but only briefly as the mechanisms involved in the degradation of mRNA are not well understood.

2. Translational control by mRNA stability

The stability of mRNA depends on its primary and secondary structure as well as on cellular factors such as stabilizing proteins and nucleases (1,2). Thus, estrogen not only induces the transcription of proteins such as vitellogenin in the liver of species that synthesize this egg protein, but at high concentrations (above physiological levels) also selectively stabilizes the mRNA (3). Withdrawal of the hormone results in cessation of synthesis as well as destabilization of the mRNA. Presumably, one or more proteins are involved in the hormone-dependent stabilization of the messenger. Similarly, the 30- to 50-fold variation in the level of histone mRNAs during the cell cycle is due largely to the destabilization of the mRNA after completion of DNA replication. This degradation of the messenger requires continued protein synthesis and probably involves recognition by free histones of a stem – loop structure at the extreme 3′-terminus of the non-polyadenylated histone mRNAs. Formation of this histone mRNA – histone complex is thought to activate a ribosome-associated 3′ – 5′ exonuclease resulting in the degradation of the mRNA (1). During S-phase, newly synthesized histones interact with the newly replicated DNA to form nucleosomes, so that the pool of free histones is kept small, thus preventing the destabilization of histone mRNAs.

3. Protein – mRNA complexes and mRNA secondary structure in translational control

3.1. Protein – mRNA complexes

Throughout the ribosome cycle, dynamic protein – mRNA interactions are functionally important in the initiation, elongation, and termination of polypeptide synthesis (see Chapter 4). In addition, more stable associations between proteins and mRNAs have been observed, particularly in eukaryotic cells. These messenger – ribonucleoprotein complexes (mRNPs) are present both in polysomes and free in the cytosol, some of the latter being either temporarily or permanently unavailable for translation (*Figure 7.1*).

Some proteins, such as the p78 poly(A)-binding protein, are present in most if not all mRNPs, others appear to be cell-specific and mRNA selective and there has been much interest in the possibility that specific differences in mRNP proteins might reflect the translational state of the mRNA. Support for this view is provided by the observation that cytosolic globin mRNP from duck reticulocytes is not translated *in vitro* whereas the mRNA obtained by deproteinizing the complex is active. Also, in unfertilized sea urchin eggs and

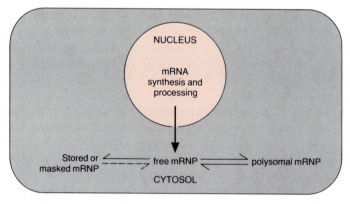

Figure 7.1. Origin of free and polysomal messenger ribonucleoproteins (mRNP). The nuclear pre-mRNA is associated with proteins which are, however, exchanged for others during transport of the processed messenger from the nucleus into the cytosol. In the cytosol, free mRNP may be bound to ribosomes and translated or be converted into stored or masked messenger which is not available for translation.

Xenopus oocytes, untranslated messenger is sequestered by association with proteins which may prevent translation until required during subsequent stages of development.

Formation of a site-specific mRNA – protein complex is involved in the translational control of the biosynthesis of ferritin, an iron storage protein which is stimulated in response to the presence of iron. In this case, a cytoplasmic repressor protein of 85 kDa binds to a highly conserved 28 nucleotide stem – loop structure in the 5'-untranslated region of ferritin mRNAs in the absence of iron but dissociates in response to iron with translational activation of the mRNAs for both the H and L subunits of ferritin (4). A similar loop motif is also involved in the translational control of transferrin receptor mRNA by an iron-responsive repressor, but in this case the relevant sequence is located in the 3'-untranslated region.

Specific regulation of gene expression at the level of translation also exists in prokaryotes. For example, the synthesis of *E.coli* threonyl – tRNA synthetase is negatively autoregulated. The mRNA for the synthetase has a tRNA-like leader structure which interacts with the synthetase thus preventing ribosome binding to the mRNA. This inhibition of protein synthesis is prevented by tRNA^{Thr} which displaces the synthetase from the mRNA. The tRNA thus acts as a translational anti-repressor and allows the cell to maintain a balance between the synthetase and its cognate tRNA.

The synthesis of some ribosomal proteins is negatively controlled by translational feedback. In this case, selective binding of one ribosomal protein to the initiation regions of the polycistronic mRNA regulates its own synthesis as well as that of other ribosomal proteins. This autogenous control may be important for ensuring a balanced synthesis of the different proteins which are required together with rRNA for the assembly of new ribosomes. Studies of the mRNA secondary structure involved in binding the ribosomal protein S4, which acts as a translational repressor of the synthesis of four ribosomal proteins (S13, S11, S4, and L17), have revealed the presence of an unusual 'double pseudoknot' linking a hairpin upstream of the ribosome binding site with sequences 2 – 10 codons downstream of the initiation codon and it has been suggested that stabilization of this structure could account for the repressor activity of S4 (6).

3.2. mRNA secondary structure

Secondary structure regulates the translation of certain kinds of eukaryotic mRNAs by a mechanism involving a ribosomal frameshift in which a directed change of the translational reading frame allows the synthesis of a single protein from two or more overlapping genes by suppression of an intervening termination codon (5). Several retroviruses (for example Rous sarcoma virus, mouse mammary tumour virus and human immunodeficiency virus type 1) use this mechanism to move from one reading frame to another to control the expression of the viral RNA-dependent DNA polymerase. Other examples where such a frame-shift operates include the synthesis of reverse transcriptase enzymes of several retrotransposons, such as the yeast Ty1, and of an avian coronavirus

(6). The mechanism of changing the reading frame appears to involve 'slippery' sequences and a complex folding of the mRNA into a structure termed a 'pseudoknot' (7).

Another interesting example is the translational control of the three cistrons of Q_β, f2 and related bacteriophage RNAs (*Figure 7.2*) which results in the

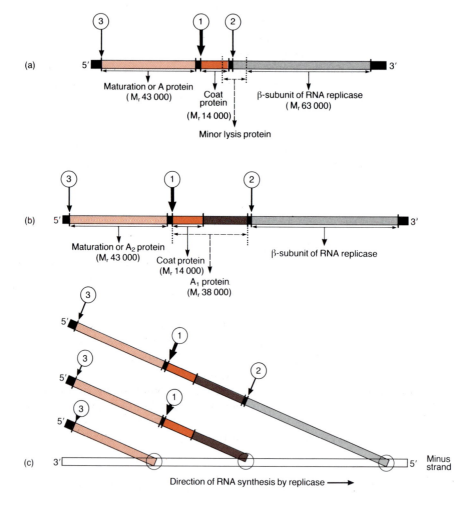

Figure 7.2. Translational control of RNA bacteriophage protein synthesis. (**a**) Arrangement of MS2 and f2 bacteriophage cistrons. (**b**) Arrangement of Q_β bacteriophage cistrons. (**c**) Replicative intermediate synthesizing new plus strands on the complementary minus strand copy of the original bacteriophage RNA. The ribosome binding sites (RBS) are indicated by numbered arrows. The major RBS (1) binds ribosomes efficiently, but can be blocked by ribosomal protein S1. The secondary RBS (2) becomes available only after translation of at least a part of the coat protein cistron by ribosomes. RBS 3 is masked by the secondary structure of native bacteriophage RNA but is accessible in nascent RNA (**c**) or *in vitro* when the secondary structure is destroyed by heating or mild treatment with formaldehyde. Non-coding regions are shown in black.

synthesis of coat protein, replicase, and A-protein in the ratio of approximately 20:5:1, both *in vivo* and *in vitro*. This quantitative control is due to differential and independent initiaton at each cistron mainly as a result of differences in the secondary structure of the initiation sites. Thus, *in vitro* protein synthesis of the replicase subunit may be increased to levels as high as that of coat protein by destruction of the secondary structure of f2 bacteriophage RNA by heating or treatment with formaldehyde. Moreover, *in vitro* synthesis of the A-protein is efficient when nascent RNA of the replicative intermediate is used as messenger presumably because the A-protein cistron has a different secondary structure in the nascent mRNA as compared with the mature RNA. It also been found that *in vivo* there is a delay in the synthesis of coat protein and this temporal control is due to translational repression of the cistron by the ribosomal protein S1. Since S1 also functions as one of the host-derived subunits of the RNA replicase, association of the newly synthesized translation product of the f2 replicase cistron with S1 would favour its dissociation from the bacteriophage RNA, thus allowing translation of the coat protein cistron to start.

3.3. Translational control involving small RNAs

Small RNAs have been detected in a variety of eukaryotic cells and are thought to be involved in modulating mRNA translation by as yet largely unknown mechanisms. Some of the RNA species stimulate *in vitro* translation whereas others inhibit, either as such or in the form of cellular ribonucleoprotein complexes. A 70–90 nucleotide inhibitory RNA which is present as a 10S RNP in the cytoplasm of chick embryonic muscle has been studied in detail. In this case, inhibition of translation may be due to competition of the RNP with the mRNA for binding to the small ribosome subunit. Amongst the stimulatory RNAs the best characterized is a small RNA, about 160 nucleotides in size, which accumulates in cells which have been infected with adenovirus. At a late stage after infection this virus-associated RNA, VA-RNA$_1$, is required to maintain general protein synthesis. It is not specific for the synthesis of viral proteins but acts by inhibiting the phosphorylation of the α-subunit of initiation factor eIF-2 (see Section 3.1) and thus maintains the recycling of guanine nucleotides which is essential for the functioning of this factor.

4. Regulation of the activity of translation factors

4.1. Control of initiation factor activity

The initiation of protein synthesis is inhibited by phosphorylation of the initiation factor eIF-2 and the mechanism of this effect has been elucidated in considerable detail. Phosphorylation is stimulated by lack of haem or the presence of double-stranded RNA. Inhibition of translation appears to result from phosphorylation of serine 51 of the α-subunit of this factor and activity can be restored by dephosphorylation. Two different protein kinases capable of phosphorylating eIF-2 have been characterized. One, called the haem-controlled repressor (HCR)

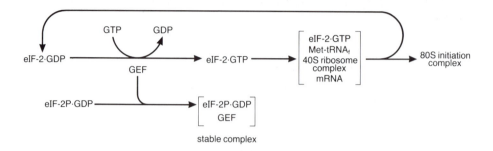

Figure 7.3. Control of initiation of mRNA translation by phosphorylation of eIF-2. Following phosphorylation of eIF-2 to eIF-2P, formation of the 80S initiation complex by the same route as shown at the top of the diagram releases eIF-2P·GDP which forms a stable complex with the guanine nucleotide exchange factor, GEF. The resulting sequestration of GEF prevents regeneration of eIF-2·GTP and thus inhibits synthesis of the initiation complex and continuation of translation.

or haem-regulated inhibitor (HRI), is a cytoplasmic protein of 95 kDa, which becomes activated by phosphorylation. The other kinase of M_r 67 kDa is activated by phosphorylation in the presence of double-stranded RNA. Thus, a cascade of protein phosphorylation is involved in this system.

Although phosphorylated eIF-2 can form an active initiation complex and initiate a round of translation, the eIF-2P·GDP complex which is released in this process is unable to exchange GDP for GTP but irreversibly binds the guanine nucleotide exchange factor (GEF) in a stable inactive complex (*Figure 7.3*). GEF is present in cells in much smaller amounts than eIF-2 and even partial phosphorylation of eIF-2 is therefore sufficient to sequester all the GEF in this complex, thus preventing regeneration of eIF-2·GTP after initiation of protein synthesis and inhibiting further rounds of translation.

Conditions other than lack of haem, for example oxidized glutathione, heat shock or serum deprivation, which are known to inhibit protein synthesis, also give rise to phosphorylation of eIF-2α. Conversely, the activity of eIF-4F is decreased by dephosphorylation of the 25 kDa subunit rather than phosphorylation and it seems that different protein kinases and phosphatases must be involved in modulating the activities of these factors.

4.2. Activity of other translation factors

There is relatively little information about modulation of elongation factor activity but a Ca^{2+}/calmodulin-dependent protein kinase has been shown recently to phosphorylate EF-2. Phosphorylated EF-2 was inactive in the poly(U)-directed synthesis of polyphenylalanine and moreover also inhibited the activity of non-phosphorylated EF-2. Dephosphorylation of the factor by phosphatase restored its activity.

5. Availability of amino acids, tRNA abundance and codon usage

Polypeptide synthesis depends on an adequate supply of all the aminoacylated tRNAs and it is not surprising that synthesis is decreased or inhibited by conditions giving rise to an imbalance in aminoacyl tRNAs or in a deficiency of any one of them.

5.1. Effects of amino acid deficiency on protein synthesis

There is evidence that amino acid deprivation reduces protein synthesis as a result of a decrease or inhibition of peptide chain elongation. Thus, O-methylthreonine, an analogue of isoleucine, partially blocks elongation at isoleucine codons and asparaginase, by removing asparagine, completely inhibits cell-free protein synthesis in reticulocyte lysates. In the latter case, the inhibition of elongation was found to be prevented or reversed by addition of asparagine (8).

5.2. Abundance of tRNA and codon usage

In the elongation cycle of protein synthesis the rate-limiting step is the search by the ternary aminoacyl-tRNA·EF-Tu·GTP complex for the cognate mRNA codon in the ribosomal A-site, the next two steps, peptide bond formation and translocation, being more rapid. Since different tRNAs are present in unequal amounts, elongation rates are slower at codons corresponding to rare tRNA species. Evidence for the existence of such translational bottlenecks has been obtained by an analysis of nascent polypeptides during the synthesis of certain proteins by *E.coli* when chains of discrete sizes were found to accumulate, indicating non-uniform rates of peptide elongation.

The existence of synonymous codons for many amino acids raises the question of preferential use of some codons and its possible significance in relation to translation efficiency and control. In bacteria, there is a divergence with respect to codon usage which is reflected in the ratio of $A + T/G + C$ in the genome. *Proteus vulgaris* has a strong preference for A or U in the third position of the codons and its genome is 62.4% $A + T$. In *Pseudomonas aeruginosa* which has a high $G + C$ content (67.2%) the most common codons are those with the strongest predicted codon–anticodon interaction and cytosine is preferred for the third codon position. It is not possible to draw any general conclusions concerning the effects of codon composition on elongation rates. Similar elongation rates have been obtained in a bacterial cell-free system with poly(UG) and poly(U) messengers, the former polymerizing cysteine and valine at a rate of 8–12 peptide bonds/sec, the latter synthesizing poly(Phe) at a rate of 10 residues/sec at 37°C. In contrast, a comparison of protein synthesis involving different codons in mutants of the *E.coli lacZ* gene showed that rare codons were translated at a rate of only 2 peptide bonds/sec, equivalent to about 20% of the rate of common codons (9). There are also many other examples indicating non-uniform translation rates at different codons.

Although codon usage rather than mRNA secondary structure is thought to determine the translation rate in *E.coli* (9) it is not clear whether such differences are of major general importance in the translational control of protein synthesis. It is possible that codon composition may be more significant in relation to the secondary structure of mRNA, which is well known to be important in the initiation of translation. Since codon degeneracy allows the mRNA coding for a specific polypeptide to have any one of many alternative nucleotide sequences, which would give rise to different secondary structures, one would expect evolution to select an mRNA structure that is favourable for efficient translation and any relevant translational control.

6. Modulation of ribosome activity

In view of the central importance of the ribosome in protein synthesis one would expect that modulation of its activity could control translation. Indeed, specific ribosomal components are known to have an important function in relation to the accuracy of protein synthesis. Thus, in *E.coli* ribosomal protein S12 determines the accuracy of codon – anticodon interactions and modulates the translational error frequency in the presence of the antibiotic streptomycin.

Whether and to what extent reversible modifications of ribosomal constituents are involved in translational regulation is unresolved. For example, phosphorylation of ribosomal protein S6 increases with cell proliferation but it is not clear whether this change enhances the translation rate or is related to some other specific function of the ribosome.

7. Further reading

Clemens,M.J. (1989) Regulatory mechanisms in translational control. *Curr. Opinion Cell Biol.*, **1**, 1160 – 7.

Dreyfuss,G. (1986) Structure and function of nuclear and cytoplasmic ribonucleoprotein particles. *Annu. Rev. Cell. Biol.*, **2**, 459 – 98.

Drysdale,J.W. (1988) Human ferritin gene expression. *Progr. Nucleic Acid Res. Mol. Biol.*, **35**, 127 – 55.

Larson,D.E. and Sells,B.H. (1987) The function of proteins that interact with mRNA. *Mol. Cell. Biochem.*, **74**, 5 – 15.

Lodish,H.F. (1976) Translational control of protein synthesis. *Annu. Rev. Biochem.*, **45**, 39 – 72.

Nomura,M., Gourse,R., and Baughman,G. (1984) Regulation of the synthesis of ribosomes and ribosomal components. *Annu. Rev. Biochem.*, **53**, 75 – 117.

Pain,V.M. (1986) Initiation of protein synthesis in mammalian cells. *Biochem. J.*, **235**, 625 – 37.

Richter,J.D. (1988) Information relay from gene to protein: the mRNP connection. *Trends Biochem. Sci.*, **13**, 483 – 6.

Richter,J.D. (1991) Translational control during early development. *BioEssays*, **13**, 179 – 83.

Sarkar,S. (1984) Translational control involving a novel cytoplasmic RNA and ribonucleoprotein. *Progr. Nucleic Acid Res. Mol. Biol.*, **31**, 267 – 93.

Starzyk,R.M. (1988) A site-specific mRNA-cytoplasmic protein complex *in vitro*. *Trends Biochem. Sci.*, **13**, 119–20.

Weissmann,C., Billeter,M.A., Goodman,H.M., Hindley,J., and Weber,H. (1973) Structure and function of phage RNA. *Annu. Rev. Biochem.*, **42**, 303–28.

Winkler,M. (1988) Translational regulation in sea urchin eggs: A complex interaction of biochemical and physiological regulatory mechanisms. *BioEssays*, **8**, 157–61.

8. References

1. Cleveland,D.W. (1989) *Curr. Opinion Cell Biol.*, **1**, 1148–53.
2.. Cleveland,D.W. (1989) *Curr. Opinion Cell Biol.*, **1**, 10–14.
3. McKenzie,E.A. and Knowland,J. (1990) *Mol. Endocrinol.*, **4**, 807–11.
4. Munro,H.N. and Eisenstein,R.S. (1989) *Curr. Opinion Cell Biol.*, **1**, 1154–9.
5. Craigen,W.J. and Caskey,C.T. (1987) *Cell*, **50**, 1–2.
6. Tang,C.K. and Draper,D.E. (1989) *Cell*, **57**, 531–6.
7. Brierley,I., Digard,P., and Inglis,S.C. (1989) *Cell*, **57**, 537–47.
8. Arnstein,H.R.V., Barwick,C.W., Lange,J.D., and Thomas,H.D.J. (1986) *FEBS Lett.*, **194**, 146–50.
9. Sørensen,M.A., Kurland,C.G., and Pedersen,S. (1989) *J. Mol. Biol.*, **207**, 365–77.

Glossary

Anticodon: three consecutive bases in tRNA that bind to a specific mRNA codon by complementary anti-parallel base-pairing.

Anti-parallel base-pairing: pairing through specific hydrogen bonds between base residues of two polynucleotide chains or two segments of a single chain with phosphodiester bonds running in the 5′ → 3′ direction in one chain or segment and in the 3′ → 5′ direction in the other. In DNA and RNA the hydrogen bonds are usually formed between complementary base pairs (A and either T or U, and G and C).

Autogenous control (autoregulation): a mechanism whereby expression of a gene is regulated by its own product.

Bihelix (double helix): the helical structure formed between two polynucleotide strands. In DNA and RNA the strands are antiparallel. A single-stranded polynucleotide, such as mRNA, has a partly bihelical structure which is formed by the molecule looping back on itself to allow anti-parallel base-pairing.

Chaperone: a molecule, usually a protein, that facilitates a biosynthetic assembly process without being incorporated into the product.

Cistron: (*see* gene).

Coding strand: the strand of the DNA double helix which serves as the template in the transcription of RNA. Its nucleotide sequence is complementary to that of the sense strand (q.v.).

Codon (coding triplet): three consecutive bases in mRNA or DNA which code for an amino acid in protein synthesis.

Cognate: having affinity; in protein synthesis it refers to an amino acid, the corresponding specific tRNA and the aminoacyl-tRNA synthetase.

C-terminal amino acid: the last amino acid in a polypeptide chain, which has a free or substituted α-carboxyl group.

Double helix: (*see* bihelix).

Downstream: on the 3′ side of mRNA or the sense strand of DNA.

Eukaryote (eucaryote): a unicellular or multicellular organism comprising cells with a specific organelle bounded by a membrane, the nucleus, that contains the chromosomal DNA.

Gene (cistron): originally defined as the unit of heredity. At the DNA level a gene is usually considered to be the region of the chromosome which codes for a single polynucleotide or polypeptide chain that is copied in the form of an RNA transcript; for example, in eukaryotes the gene sequence for a single polypeptide chain usually extends from the start of transcription through the coding region and intervening sequences to the polyadenylation site.

Initiation codon: a group of three consecutive nucleotides, usually AUG, at the beginning of the mRNA coding sequence which determines the reading frame and signals the start of polypeptide chain synthesis.

Initiation sequence: a sequence of nucleotides on the 5′ side of the mRNA initiation codon (including in prokaryotes the Shine–Dalgarno sequence) that is involved in binding to the small ribosomal subunit.

Intercistronic: the nucleotide sequence(s) between individual cistrons in mRNA or DNA.

Isosteric: pertaining to compounds or groups that are similar in physical properties and have a similar number and arrangement of valence electrons.

mRNA cap: the 7-methylguanylic acid residue which is linked by a pyrophosphate group to the 5′ terminus of eukaryotic messenger RNA.

Nascent: growing, developing, e.g. the nascent polypeptide chain which is an intermediate in the synthesis of the polypeptide.

N-terminal amino acid: the first amino acid of a polypeptide chain, which may have a free or substituted α-amino group.

Polarity: the asymmetry of a polymer such as a polynucleotide or polypeptide. In DNA the two strands have opposite polarity, i.e. they run in opposite directions ($5′ \rightarrow 3′$ and $3′ \rightarrow 5′$). The polarity of a polypeptide is defined as running from the N-terminus to the C-terminus.

Polycistronic: (*see* cistron).

Polyprotein: a protein precursor which is synthesized by translation of a single mRNA molecule and processed by proteolytic cleavage into two or more functional polypeptides.

Prokaryote (procaryote): a unicellular organism lacking a nuclear membrane and containing chromosomal DNA in the cytoplasm.

Prosthetic group: a low-molecular-weight non-protein molecule strongly bound to a protein and essential for its function.

Reading frame: the selection of one of three possible ways of translating groups of three consecutive nucleotides in mRNA. It is determined by the initiation codon.

Read-through protein: a protein that is produced as a result of a failure to terminate translation at the end of the coding sequence, thus containing additional amino acid residues in the polypeptide chain compared with the normal protein.

Retrotransposon: a mobile genetic element that transposes *via* an RNA intermediate.

Sense strand: the strand of the DNA double helix which has the same nucleotide sequence (after removing introns and substituting U for T) as mRNA. It is also known as the anticoding or non-coding strand.

Sequence context: the relationship of neighbouring or distant residues to the function of a particular sequence in a biopolymer.

Signal sequence: a nucleotide sequence present in mRNA and the sense strand of DNA including the initiation codon which codes for the signal peptide of the precursor of secretory proteins.

Subunit: 1. in a multimeric protein, the smallest covalent unit, which may consist of a single polypeptide chain or of more than one, covalently linked polypeptides. 2. a definite substructure of a complex entity such as the ribosome, which comprises a small and a large subunit, each containing both RNA and protein.

Suppressor tRNA: an unusual transfer RNA, produced by a suppressor gene, capable of interaction with a termination codon that has been introduced into mRNA by a mutation. In this way suppressor tRNA prevents premature termination of translation which would otherwise occur.

Translation: synthesis of a polypeptide with an amino acid sequence determined by the coding sequence of mRNA.

Translocation: 1. in protein synthesis, the reaction by which peptidyl tRNA is transferred from the A to the P site of the ribosome. 2. the transfer of a molecule across a membrane. 3. the insertion of a chromosome fragment into another, non-homologous chromosome.

Upstream: on the 5′ side of mRNA of the sense strand of DNA.

Index

IN FOCUS®

The 'In Focus' series is specifically written for students facing the problem of keeping up to date with fast moving areas of biology and medicine. Each title presents the very latest information in a clear and accessible format, providing an in-depth knowledge of the subject.

Proteins are of central importance to the structure and function of living cells. The topic of the biosynthesis of these molecules is thus central to any study of biochemistry: it is a complex process utilizing many components and much cellular energy. This book describes the principal mechanisms involved, with particular emphasis on recent investigations into the contributions of transfer RNA, messenger RNA, protein factors, and ribosomes to peptide bond formation. Also discussed are: modification of amino acid side chains ● protein secretion and relocation mechanisms ● control of protein synthesis via mRNA translation.

ISBN 0-19-963040-2

9 780199 630400

O9-BTD-628

"Stimulating the imagination is not an alternative educational activity to be argued for in competition with other claims; it is a prerequisite to making any activity educational."

In the current era of prescribed objectives, testing, and technical emphases in the curriculum, *Imagination and Education* is unique in its championing of a positive, liberating concept of imagination in education.

The contributors to this volume strongly demonstrate the vital connections between imagination and education in a way that specifically addresses the classroom experience. Combining conceptual clarity with practical insights from diverse fields such as developmental psychology, science, art, music, philosophy, and literature, the chapters explore systematically the specific role of the imagination in the school. The examples given throughout not only establish logical links between imagination and education, but also furnish a collective vision of what a full, creative education is all about.

Noted chapter authors including Maxine Greene, Brian Sutton-Smith, Roger Shepard, Robin Barrow, and British Poet Laureate Ted Hughes combine serious scholarship with an engaging enthusiasm about their subjects.

This timely book will be essential for students in educational and curriculum theory, aesthetic and art education, psychology, and teacher training programs. As a professional text for teacher educators, it is superb, and will be of interest to every thoughtful person who cares about reconceiving and revitalizing the education of young people today.

Kieran Egan is Professor of education at Simon Fraser University in Canada. His books include *Education and Psychology* and *Teaching as Story Telling*.

Dan Nadaner is a painter who has taught art at the elementary, secondary, and college levels. Formerly Assistant Professor, Faculty of Education, Simon Fraser University, he has published numerous articles on art and education.

ALSO OF INTEREST—

Awakening the Inner Eye
Intuition in Education
Nel Noddings and Paul J. Shore
1984/Paper/ISBN 0-8077-2899-3

Landscapes of Learning
Maxine Greene
1978/Paper/ISBN 0-8077-2534-X

Cover design by Paul Lansdale

Teachers College
Columbia University
New York, N.Y. 10027

ISBN 0-8077-2877-2